Mind Children

MIND CHILDREN

*The Future of Robot
and Human Intelligence*

Hans Moravec

HARVARD UNIVERSITY PRESS
Cambridge, Massachusetts
London, England   1988

This book is printed on acid-free paper,
and its binding materials have been
choosen for strength and durability.

Library of Congress Cataloging-in-Publication Data available
88-21343
ISBN 0-674-57616-0

In memory of my father, who taught me to tinker

To my mother, who taught me to read

To Ella, who made me complete

Contents

Figures

Mind Children

Prologue

ENGAGED for billions of years in a relentless, spiraling arms race with one another, our genes have finally outsmarted themselves. They have produced a weapon so powerful it will vanquish the losers and winners alike. This device is not the hydrogen bomb—widespread use of nuclear weapons would merely delay the immensely more interesting demise that has been engineered.

What awaits is not oblivion but rather a future which, from our present vantage point, is best described by the words "postbiological" or even "supernatural." It is a world in which the human race has been swept away by the tide of cultural change, usurped by its own artificial progeny. The ultimate consequences are unknown, though many intermediate steps are not only predictable but have already been taken. Today, our machines are still simple creations, requiring the parental care and hovering attention of any newborn, hardly worthy of the word "intelligent." But within the next century they will mature into entities as complex as ourselves, and eventually into something transcending everything we know—in whom we can take pride when they refer to themselves as our descendants.

Unleashed from the plodding pace of biological evolution, the children of our minds will be free to grow to confront immense and fundamental challenges in the larger universe. We humans will benefit for a time from their labors, but sooner or later, like natural children, they will seek their own fortunes while we, their aged parents, silently fade away. Very little need be lost in this passing of the torch—it will be in our artificial offspring's power, and to their benefit, to remember almost everything about us, even, perhaps, the detailed workings of individual human minds.

The process began about 100 million years ago, when certain gene lines hit upon a way to make animals with the ability to learn some behaviors from their elders during life, rather than inheriting them all at conception. It was compounded 10 million years ago when our primate ancestors began to rely on tools made of bones, sticks, and stone, and accelerated again with the harnessing of fire and the development of complex languages about 1 million years ago. By the time our species appeared, around 100 thousand years ago, cultural evolution, the juggernaut our genes had unwittingly constructed, was rolling with irresistible momentum.

Within the last 10 thousand years, changes within the human gene pool have been inconsequential in comparison with the snowballing advances in human culture. We have witnessed first an agricultural revolution, followed by the establishment of large-scale bureaucratic governments with the power to levy taxes for their support, the development of written languages, and the rise of leisure classes with time and energy to devote to intellectual concerns. In the last thousand years or so, inventions beginning with movable type printing have greatly speeded the flow of cultural information, and thus its evolutionary pace.

With the coming of the industrial revolution 200 years ago, we entered the final phase, one in which artificial substitutes for human body functions such as lifting and transporting have become ever more economically attractive—indeed, indispensable. Then, 100 years ago, with the invention of practical calculating machines, we were able for the first time to artificially duplicate some small but vexing functions of the human mind. The computational power of mechanical devices has risen a thousandfold every 20 years since then.

We are very near to the time when virtually no essential human function, physical or mental, will lack an artificial counterpart. The embodiment of this convergence of cultural developments will be the intelligent robot, a machine that can think and act as a human, however inhuman it may be in physical or mental detail. Such machines could carry on our cultural evolution, including their own construction and increasingly rapid self-improvement, without us, and without the genes that built us. When that happens, our DNA will find itself out of a job, having lost the evolutionary race to a new kind of competition.

A. G. Cairns-Smith, a chemist who has contemplated the beginnings of life on the early earth, calls this kind of internal coup a *genetic takeover*. He suggests that it has happened at least once before. In *Seven Clues to the Origin of Life*, Cairns-Smith argues that the precursors to life as we know it were microscopic crystals of clay, which reproduced by the simple process of crystal growth. Most crystals are marked by patterns of dislocation in the orderly arrangement of their atoms, many of which propagate as the crystal grows. If the crystal should fracture, each piece may inherit a copy of the pattern, sometimes with a slight change. Such defects can have a dramatic effect on a clay's physical and chemical properties. Crystals sharing one dislocation pattern may form dense clumps, while those with another may aggregate into a spongy mass. Mineral-bearing water may be diverted around one type but trickle through the other, providing raw materials for continued growth. The patterns also affect growth indirectly by modulating the chemistry of other molecules in their environment. Clays are powerful chemical catalysts; the tiny crystals have enormous total surface area, to which molecules can adhere in certain configurations, depending on the external shape of the crystal and molecule in question. These common crystals thus possess the essentials for Darwinian evolution—reproduction, inheritance, mutation, and differences in reproductive success.

In Cairns-Smith's theory, the first genetic takeover began when some clay species, in vigorous Darwinian competition with one another, began to encode some genetic information externally in long carbon molecules. Such polymers are more stable than the easily disturbed dislocation patterns themselves, and organisms using them to ever greater extent reproduced more successfully. Although utterly dependent at first on the existing crystal-based chemical machinery, as these carbon molecules assumed a greater share of the reproductive role they became less reliant on the crystals. In time, the simple crystal scaffolding vanished altogether, leaving in its evolutionary wake the complex, interdependent system of organic machinery we call life.

Today, billions of years later, another change is under way in how information passes from generation to generation. Humans evolved from organisms defined almost totally by their organic genes. We now rely additionally on a vast and rapidly growing corpus of cultural information generated and stored outside our genes—in our nervous systems, libraries, and, most recently, computers. Our culture still depends utterly on biological human beings, but with each passing year our machines, a major product of the culture, assume a greater role in its maintenance and continued growth. Sooner or later our machines will become knowledgeable enough to handle their own maintenance, reproduction, and self-improvement without help. When this happens, the new genetic takeover will be complete. Our culture will then be able to evolve independently of human biology and its limitations, passing instead directly from generation to generation of ever more capable intelligent machinery.

Our biological genes, and the flesh and blood bodies they build, will play a rapidly diminishing role in this new regime. But will our minds, where culture originated, also be lost in the coup? Perhaps not. The coming revolution may liberate human minds as effectively as it liberates human culture. In the present condition we are uncomfortable halfbreeds, part biology, part culture, with many of our biological traits out of step with the inventions of our minds. Our minds and genes may share many common goals during life, but there is a tension between time and energy spent acquiring, developing, and spreading ideas and effort expended toward maintaining our bodies and producing a new generation (as any parent of teenagers can observe). The uneasy truce between mind and body breaks down completely as life ends. Our genes usually survive our death, grouped in different ways in our offspring and our relatives. In a subtle way it is no doubt in their evolutionary interest to regularly experiment like this with fresh shuffles of the genetic deck. But the process is devastating for our other half. Too many hard-earned aspects of our mental existence simply die with us.

It is easy to imagine human thought freed from bondage to a mortal body—belief in an afterlife is common. But it is not necessary to adopt a mystical or religious stance to accept the possibility. Computers provide a model for even the most ardent mechanist. A computation in progress—what we can reasonably call a computer's thought process—can be halted in midstep and transferred, as program and data read out of the machine's memory, into a physically different computer, there to resume as though nothing had happened. Imagine that a human mind might be freed from its brain in some analogous (if much more technically challenging) way.

A mind would require many modifications to operate effectively after being rescued from the limitations of a mortal body. Natural human mentality is tuned for a life span's progression from impressionable plasticity to self-assured rigidity, and thus is unpromising material for immortality. It would have to be reprogrammed for continual adaptability to be long viable. Whereas a transient mortal organism can leave the task of adaptation to the external processes of mutation and natural selection, a mind that aspires to immortality, whether it traces its beginnings to a mortal human being or is a completely artificial creation, must be prepared to adapt constantly from the inside.

Perhaps it would undergo a cyclical rejuvenation, acquiring new hardware and software in periodic phases that resemble childhood. Or maybe it could update the contents of its mind and body continuously, adding and deleting, testing components in all kinds of combinations, to keep up with changing conditions. The testing is of central importance: it steers the evolution. If the individual makes too many bad decisions in these tests, it will fail totally, in the old-fashioned Darwinian way.

A postbiological world dominated by self-improving, thinking machines would be as different from our own world of living things as this world is different from the lifeless chemistry that preceded it. A population consisting of unfettered mind children is quite unimaginable. We are going to try to imagine some of the consequences anyway.

1 | *Mind in Motion*

I BELIEVE that robots with human intelligence will be common within fifty years. By comparison, the best of today's machines have minds more like those of insects than humans. Yet this performance itself represents a giant leap forward in just a few decades.

Mechanical imitations of certain human functions have been with us for centuries. Many medieval clock towers are equipped with mechanisms that mark the hours with elaborate morality plays enacted by mechanical saints, knights, bishops, angels, demons, and all kinds of animals. Smaller devices that walked, talked, swam, breathed, ate, wrote with quill pens, or played musical instruments have amused polite society since at least the fifteenth century. Leonardo da Vinci, for one, constructed elaborate mechanical displays of this sort for his patrons. These early clockwork machines—powered by running water, falling weights, or springs—copied the motions of living things, but they could not respond to the world around them. They could only *act*, however charmingly.

Electrical, electronic, and radio technology, developed early in this century, made possible machines that could *react*—to light, sound, and invisible remote control. The result was a number of entertaining demonstration robots—as well as thoughts and stories about future humanlike machines. But only simple connections between the sensors and motors were possible at first. These newer machines could sense as well as act, but they could not *think*.

Machines That Think (Weakly)

During World War II analog computers—machines that simulated physical systems by representing their changing quantities as analogous moves of shafts or voltages—were designed for controlling anti-aircraft guns, for navigating, and for precision bombing. Some of their developers noticed a similarity between the operation of these devices and the regulatory systems in living things, and these researchers were inspired to build machines that acted as though they were alive. Norbert Wiener at the Massachusetts Institute of Technology (MIT) coined the term *cybernetics* for this unified study of control and communication in animals and machines. Its practitioners combined new theory on feedback regulation with advances in postwar electronics and early knowledge of living nervous systems to build machines that were able to respond like simple animals and to learn. The rudiments of artificial thought had arrived.

Among the highlights of the cybernetics effort was a series of electronic turtles built during the 1950s by W. Grey Walter, a British psychologist. With subminiature radio-tube electronic brains, rotating phototube eyes, microphone ears, and contact-switch feelers, the first versions could locate their recharging hutch when their batteries ran low and otherwise avoid trouble while wandering about. Groups of them exhibited complex social behavior by responding to one another's control lights and touches. A later machine with the same senses could be conditioned to associate one stimulus with another and could learn by repeated experience that, for instance, a loud noise would be followed by a kick to its shell. Once educated, the turtle would avoid a noise as it had before responded to a kick. The associations were slowly accumulated as electrical charges in electronic devices called *capacitors*, used here as memory devices.

Perhaps the most impressive creation of the cyberneticists was the Johns Hopkins Beast. Built by a group of brain researchers in the early 1960s, it wandered the halls guided by sonar and a specialized photocell eye that searched for the distinctive black cover plate of wall outlets, where it would plug itself in to feed. The Beast inspired a number of imitators. Some used special circuits connected to television cameras instead of photocells and were controlled by assemblies of (then new) transistor digital logic gates, like those that can now be found, in thousands and millions, in the integrated circuits of every computer. Some added new motions such as "Shake to untangle recharging arm" to the repertoire of basic actions.

The field of cybernetics thrived less than two decades. As is so often the case, it was eclipsed by a relative, the artificial intelligence movement. The war's many small analog computers, which had inspired cybernetics, had a few much larger digital cousins. These machines computed not by the measured turns of shafts or flow of current but by counting, in discrete jumps. The first automatic digital computers—huge, immobile, autonomous calculators—were completed toward the end of the war. Colossus, an ultrasecret British machine that broke the German Enigma code and helped to change the course of the war, scanned code keys tens of thousands of times faster than humanly possible. In the United States, ENIAC computed antiaircraft artillery tables for the Army and later did calculations for the construction of the atomic bomb, at speeds similar to Colossus.

Less belligerently, these "giant brains," as they came to be called, provided unprecedented opportunities for experiments in complexity. Pioneers like Alan Turing, one of the creators of Colossus, and John von Neumann, who was involved with the first American machines, harbored the hope that the ability to think rationally, our unique asset in dealing with the world, could be captured in a machine. Our minds might be amplified by computers just as our muscles had been amplified by the steam engines of the industrial revolution. Programs to reason and to play intellectual games like chess were designed by Claude Shannon of MIT and by Turing in 1950, but the earliest computers were too limited and expensive for this use. A few poor checker-playing programs did appear on the first commercial machines of the early 1950s, and equally poor chess programs showed up in the last half of that decade, along with a good checker player by Arthur Samuel of IBM. Then in 1957 Allen Newell and Herbert Simon of Carnegie Tech (now Carnegie Mellon University) and John Shaw of the RAND Corporation demonstrated the Logic Theorist, the first program able to reason about arbitrary matters by starting with axioms and applying rules of inference to prove theorems.

In 1960 John McCarthy, then at MIT, coined the term *artificial intelligence* (AI) for the effort to make computers think. By 1965 the first students of McCarthy, Marvin Minsky (also at MIT), Newell, and Simon had produced AI programs that proved theorems in geometry, solved problems from intelligence tests, algebra books, and calculus exams, and played chess, all with the proficiency of an average college freshman. Each program could handle only one narrow type of problem, but for first efforts these programs were encouraging—so encouraging that most people involved felt that another decade of progress would surely produce a genuinely intelligent machine. This was an understandable miscalculation.

Now, a quarter of a century later, computers are thousands of times more powerful than these sixties models, but they do not seem much smarter. By 1975 progress in artificial intelligence had slowed from the heady sprint of a handful of enthusiasts to the plodding trudge of growing throngs of workers. Even so, modest successes have maintained flickering hope. So-called *expert systems,* programs encoding the decision rules of human experts in narrow domains such as diagnosis of disease, factory scheduling, or computer system configuration, are daily earning their keep in the business world. A fifteen-year effort at MIT gathered knowledge about algebra, trigonometry, calculus, and related fields into a wonderful program called MACSYMA, now marketed commercially, that manipulates symbolic formulas and helps to solve otherwise forbidding problems. Several chess-playing programs are now officially rated as chess masters, and excellent performance has been achieved in other games like backgammon. There are semi-intelligent programs that can understand simplified typewritten English about restricted subjects and make elementary deductions in the course of answering questions about these texts. Some interpret spoken commands chosen from thousand-word repertoires, and others can do simple visual tasks, such as deciding whether a part is in its desired location.

Unfortunately for humanlike robots, computers are at their worst trying to do the things most natural to humans, such as seeing, hearing, manipulating objects, learning languages, and commonsense reasoning. This dichotomy—machines doing well things humans find hard, while doing poorly what is easy for us—is a giant clue to the problem of how to construct an intelligent machine.

Machines That See (Dimly) and Grasp (Clumsily)

In the mid-1960s Marvin Minsky's students at MIT began to connect television cameras and mechanical robot arms to their computers, giving eyes and hands to artificial minds so that their machines could see, plan, and act. By 1965 these researchers had created a machine that could find and remove white blocks from a black tabletop. This accomplishment required a controlling program as complex as any of the then-current pure reasoning programs—programs which, unencumbered by robot appendages, could, for instance, match first-year college students in solving calculus problems. Yet Minsky's hand–eye system could be bested by a toddler. Nevertheless, the experiments continued at MIT and elsewhere, gradually developing into a field which now goes by the name *robotics*, a term coined in 1942 in a science fiction story by Isaac Asimov from the word *robot*, itself coined by the Czech playwright Karel Capek in 1921. Robotics started far lower on the scale of human performance than artificial intelligence, but its progress in the past twenty years has been just as agonizingly slow and difficult.

Not all robots, nor all people, idle away their lives in universities. Many must work for a living. Even before the industrial revolution, before any kind of thought was mechanized, partially automatic machinery, powered by wind or flowing water, was put to work grinding grain and cutting lumber. The beginning of the industrial revolution in the eighteenth century was marked by the invention of a plethora of devices that could substitute for manual labor in a precise, and thoroughly inhuman, way. Driven by shafts turned by water or steam, these machines pumped, pounded, cut, spun, wove, stamped, moved materials and parts, and much else, consistently and tirelessly.

Once in a while something ingeniously different appeared. For instance, the Jacquard loom, invented in 1801, could weave intricate tapestries specified by a string of punched cards. By the early twentieth century, electronics had given industrial machines limited senses; they could now stop when something went wrong, or control the temperature, thickness, even consistency of their workpieces. Still, each machine did one job and one job only. Consequently, the product produced by the machine often became obsolete before the machine had paid back its design and construction costs. This problem had become particularly acute by the end of World War II.

In 1954 the inventor George Devol filed a patent for a new kind of industrial machine, the programmable robot arm, whose task could be altered simply by changing the stream of punched program cards that controlled its movement. In 1958, with Joseph Engelberger, Devol founded a company named Unimation (a contraction of "universal" and "automation") to build such machines. The punched cards soon gave way to magnetic memory, which allowed the robot to be programmed simply by leading it once by the hand, so to speak, through its required paces. The first industrial robot began work in a General Motors plant in 1961. To this day, most large robots that weld, spray paint, and move pieces of cars are of this type.

Only when the cost of small computers dropped to less than $10,000 did robotics research at the universities begin to influence the design of industrial robots. The first industrial vision systems, able to locate and identify parts on conveyor belts, and usually coupled with a new class of small robot arms, appeared in the late 1970s. Robots able to see and feel, after a fashion, now play a modest but quietly booming role in the assembly and inspection of small devices like calculators, printed circuit boards, typewriters, and automobile water pumps. Indeed, industrial needs have strongly influenced university research. What was once a negligible number of smart robot projects has swelled to the hundreds on campuses across the country. And while cybernetics may now be relatively dormant, its stodgy parent, control theory, has been quite active since the war in an effort to meet the profitable needs of the aerospace industry. Elaborate methods developed to control aircraft, spacecraft, and weapons are now influencing the design of industrial robots.

In 1987 I was treated to a tour of the factory in Fremont, California, where Apple's Macintosh computers are assembled. I found most of the plant well organized but unremarkable. Many assembly steps were done manually. The most efficient machines were probably those that inserted components into circuit boards. Acting something like sewing machines, these "board stuffers" take components strung on tapes like machine-gun ammunition and "stitch" them into printed circuit boards at blinding speed, several components per second, with the board sliding rapidly into position for each part, front and back, left and right. The machines are marvels of computerized control, and very cost-effective for high-volume production, but they left me vaguely disappointed. In one small niche, however, I saw a quite different device inserting components that the high-speed machines could not handle. The parts were old-fashioned inductors—small metal cans containing a coil of wire. Each can had metal tabs that were to fit into slots in the board, and the coils ended in wires intended for small holes. Unlike the precisely shaped components on the feed tapes of the other machines, the inductors, supplied neatly arrayed on tex-tured plastic trays, often had slightly bent tabs and wires that would simply be mangled by a blind attempt to push them into a board.

The insertion machine worked in a glass booth. Boards and trays of inductors arrived and left by conveyor belts. The insertion process began with a TV camera that looked down on the parts tray. A vision program located an inductor, and a fast robot arm swooped to pick it up and then brought it, wire and tab end up, in front of a second TV camera. A second vision program examined the leads and, if they were straight enough, directed the arm to insert the component in the board. If the leads were slightly bent, the inductor was first pushed against a "pin straightener," a metal block with tapered holes, after which the pins would be inspected again. If the leads were hopelessly mangled, the inductor was dropped into a reject bin and another was fetched from the tray.

The insertion was itself a sensitive process. The inductor was rapidly brought to within a few millimeters of the board surface, then slowly lowered until the robot arm encountered resistance. The arm nudged the inductor to and fro while maintaining a slight downward pressure, until the tabs and wires found their holes, at which point it applied greater pressure to seat the component. A motorized cutter mounted below the board then cut and bent the protruding metal, anchoring the inductor. If the attempt to seat the part failed after a few seconds, it would again be brought in front of the lead-checking camera, and possibly into the pin straightener, before another insertion attempt. If a third attempt failed, the part would be tossed into the reject bin and a new one tried.

All this happened very rapidly—a part could be inserted every three or four seconds, though a troublesome one might take up to ten. I was impressed—and a little nostalgic. The activities had a familiar feel. I had been a regular witness to somewhat slower and clumsier versions of them a decade earlier at the Stanford Artificial Intelligence Lab, where I was a graduate student. In fact, Apple's assembly system was a product of a small southern California company called Adept that can trace its ancestry back to PhD theses at SAIL. The seeds cast there were starting to sprout.

The goal of humanlike performance by stationary robots, though highly diluted by a myriad of approaches and short-term goals, has acquired a relentless, Darwinian vigor. As a story, it becomes bewildering in its diversity and interrelatedness. Let us move on to the sparser world of robots that rove.

Machines That Explore (Haltingly)

The first reasoning programs needed very little data to do their work. A chessboard, or a problem in logic, geometry, or algebra, could be described by a few hundred well-chosen symbols. Similarly, the rules for solving the problem could be expressed as several hundred so-called "transformations" of this data. The difficulty lay only in finding a sequence of transformations that solved the problem, from among the astronomical number of possible combinations. It was obvious that solving problems in less restricted areas (the question "How can I get to Timbuktu?" was an often-used rhetorical example) would require a much greater initial store of information. It seemed unlikely that all the facts needed to solve such problems could be provided manually to reasoning programs.

Some facts might be made available if the programs could be taught to read and understand books, but comprehending even simple words would require detailed knowledge of the physical world. Such knowledge is assumed to preexist in the minds of book readers—no book attempts a comprehensive definition of a rock, a tree, the sky, or a human being. Possibly some of this *world knowledge*, as it has come to be called, could be obtained by the machine itself if it could directly observe its surroundings through camera eyes, microphone ears, and feeling robot hands. The desire to automate the acquisition of world knowledge was one of the early rationales for robotics projects in the artificial intelligence labs. The internal model of the world that these computers might develop could then be used by them, and by other machines, as a basis for commonsense reasoning.

Although a machine that can move around should be able to gather much more data than an immobile one, the logistical difficulty of connecting a huge immobile computer to a complicated array of sensors on a moving platform made fixed hand–eye systems more attractive to most researchers. Besides, it was soon realized that the problem of systematically acquiring information from the environment was much less tractable than the mental activities the information was intended to serve. During the 1970s dozens of research labs had robot arms connected to computers, but hardly any had robot vehicles. Most robotics researchers viewed mobility as an unnecessary complication to an already overly difficult problem. Their experience was in marked contrast with the attitude of the cyberneticists (and hundreds of hobbyists and toymakers), who had been quite satisfied with eliciting simple animal-like behavior from the modest circuitry aboard their many small mobile machines.

Stanford Research Institute's Shakey, completed in 1969, was the first, and is still the only, mobile robot to be controlled primarily by programs that reasoned. It is an exception that proves the rule. Shakey's instigators—Nils Nilsson, Charles Rosen, and others—were inspired by the early success in artificial intelligence research. They sought to apply logic-based problem-solving methods to a real-world task involving a machine that could move and sense its environment. The problems of controlling this movement and interpreting sensory data were of secondary importance to the designers, however, whose main interest was in the machine's ability to reason. The job of developing the mobility and sensory software was relegated to junior programmers.

Five feet tall and driven by two slow but precise stepping motors, Shakey was equipped with a television camera and was remote-controlled by a large computer. Methods from MIT's blocks manipu-lating programs, previously used only with robot arms, were adapted for interpreting the TV images. These worked only when the scene consisted solely of simple, uniformly colored, flat-faced objects, so a special environment was constructed for the robot. It consisted of several rooms bounded by clean walls, containing a number of large, uniformly painted blocks and wedges. Shakey's most impressive performance, executed piecemeal over a period of days, was to solve a so-called "monkeys and bananas" problem. Told to push a particular block that happened to be resting on a larger one, the robot constructed and acted on a plan that included finding a wedge that could serve as a ramp, pushing it against the large block, driving up the ramp, and delivering the requested push.

The environment and the problem were contrived, but they pro-vided a motivation, and a test, for a clever reasoning program called STRIPS (the STanford Research Institute Problem Solver). Given a task for the robot, STRIPS assembled a plan out of the limited actions the robot could take, each of which had preconditions (for example, to push a block, it must be in front of me) and probable consequences (after I push a block, it is moved). The state of the robot's world was represented in sentences of mathematical logic, and formulating a plan was like proving a theorem, the initial state of the world being the axioms, primitive actions being the rules of inference, and the desired outcome playing the role of the theorem. One complication was immediately evident: the outcome of an action is not always what one expects (as when the block does not budge). Shakey had a limited ability to handle such glitches by occasionally observing parts of the world and adjusting its internal description and replanning its actions if the conditions were not as it had assumed.

Shakey was impressive in concept but pitiable in action. Each move of the robot, each glimpse taken by its camera, consumed about an hour of computer time and had a high likelihood of failure. The block-pushing exercise described above was staged, and filmed, in steps, with several steps requiring repeated "takes" before they succeeded. The fault lay not in the STRIPS planner, which produced good plans when given a good description of what was around the robot, but in the programs that interpreted the raw data from the sensors and acted on the recommendations.

It seemed to me, in the early 1970s, that some of the creators of successful reasoning programs suspected that the poor performance in the robotics work somehow reflected the intellectual abilities of those attracted to that side of the research. Such intellectual snobbery is not unheard of, for instance between theorists and experimentalists in physics. But as the number of demonstrations has mounted, it has become clear that it is comparatively easy to make computers exhibit adult-level performance in solving problems on intelligence tests or playing checkers, and difficult or impossible to give them the skills of a one-year-old when it comes to perception and mobility.

In hindsight, this dichotomy is not surprising. Since the first multi-celled animals appeared about a billion years ago, survival in the fierce competition over such limited resources as space, food, or mates has often been awarded to the animal that could most quickly produce a correct action from inconclusive perceptions. Encoded in the large, highly evolved sensory and motor portions of the human brain is a billion years of experience about the nature of the world and how to survive in it. The deliberate process we call reasoning is, I believe, the thinnest veneer of human thought, effective only because it is supported by this much older and much more powerful, though usually unconscious, sensorimotor knowledge. We are all prodigious olympians in perceptual and motor areas, so good that we make the difficult look easy. Abstract thought, though, is a new trick, perhaps less than 100 thousand years old. We have not yet mastered it. It is not all that intrinsically difficult; it just seems so when we do it.

Organisms that lack the ability to perceive and explore their environment do not seem to acquire anything that we would call intelligence. We need make only the grossest comparison of the plant and animal kingdoms to appreciate the fact that mobile organisms tend to evolve the mental characteristics we associate with intelligence, while immobile ones do not. Plants are awesomely effective in their own right, but they have no apparent inclinations toward intelligence. Perhaps, given much more time, an intelligent plant could evolve—some carniverous and "sensitive" plants show that something akin to nervous action is possible—but the life expectancy of the universe may be insufficient time.

The cybernetics researchers, whose self-contained experiments were often animal-like and mobile, began their investigation of nervous systems by attempting to duplicate the sensorimotor abilities of animals. The artificial intelligence community ignored this approach in their early work and instead set their sights directly on the intellectual acme of human thought, in experiments running on large, stationary mainframe computers dedicated to mechanizing pure reasoning. This "top-down" route to machine intelligence made impressive strides at first but has produced disappointingly few fundamental gains in over a decade. While cybernetics scratched the underside of real intelligence, artificial intelligence scratched the topside. The interior bulk of the problem remains inviolate.

All attempts to achieve intelligence in machines have imitated natural intelligence, but the different approaches have mimicked different aspects of the original. Traditional artificial intelligence attempts to copy the conscious mental processes of human beings doing particular tasks. Its limitation is that the most powerful aspects of thought are unconscious, inaccessible to mental introspection, and thus difficult to set down formally. Some of the cyberneticists, taking a different tack, had focused on building models of animal nervous systems at the neural level. This approach is limited by the astronomical number of cells in large nervous systems and by the great difficulty of determining exactly what individual neurons do, how they are interconnected, and what nerve networks do. Both traditional AI and neural modeling have contributed insights to the enterprise, and no doubt each could solve the whole problem, given enough time. But with the present state of the art, I feel that the fastest progress can be made by imitating the *evolution* of animal minds, by striving to add capabilities to machines a few at a time, so that the resulting sequence of machine behaviors resembles the capabilities of animals with increasingly complex nervous systems. A key feature of this approach is that the complexity of these incremental advances can be tailored to make best use of the problem-solving abilities of the human researchers and the computers involved. Our intelligence, as a tool, should allow us to follow the path to intelligence, as a goal, in bigger strides than those originally taken by the awesomely patient, but blind, processes of Darwinian evolution.

The route is from the bottom up, and the first problems are those of perception and mobility, because it is on this sensorimotor bedrock that human intelligence developed. Programs which tackle incremental problems similar to those that faced early animals—how to deal with, and even to anticipate, the sudden surprises, dangers, and opportunities encountered by an exploring organism—are being written and tested in robots that have to face the uncertainties of a real world. Most approaches will fail, but a few will succeed, in much the same way that a tiny fraction of the spontaneous mutations that appear in organisms survive into the next generation. Survival depends on the advantages a new technique offers in coping with the challenges of a complex and dynamic environment. By setting up experimental conditions analogous to those encountered by animals in the course of evolution, we hope to retrace the steps by which human intelligence evolved. That animals started with small nervous systems gives confidence that today's small computers can emulate the first steps toward humanlike performance. Where possible, our efforts to simulate intelligence from the bottom up will be helped by biological peeks at the "back of the book"—at the neuronal, morphological, and behavioral features of animals and humans, as revealed by the people who study those aspects of life.

The modern robotics effort is just twenty years old, and for only the last ten of those have computers been routinely available to control robots. The recapitulation of the evolution of intelligent life is at a very early stage—robotic equivalents of nervous systems exist, but they are comparable in complexity to the nervous systems of worms. Nevertheless, the evolutionary pressures that shaped life are already

Intelligence on Earth

Post-DNA reproduction

Culture
Tools

Learning
Warm blood

Walkers
Brains
Neurons

Death

Multicelled animals
Free oxygen
Sex

Nucleated cells

DNA

Pre-DNA reproduction

Mammals

Reptiles

Birds

Amphibians

Fish

Vertebrates

Enchinoderms

Animals

Cephalopods

Bivalves

Mollusks

Arthropods

Plants

Blue-green algae

Chloroplasts

Mitochondria

Fungi   Bacteria

Earliest cells

The dawn of life

Present

100 million

500 million

1 billion

2 billion

3 billion

4 billion years ago

This partial family tree of terrestrial organisms suggests the linkage between mobility and intelligence. One and a half billion years ago our unicelled ancestors parted genetic company with the plants. As single cells, both lines were free swimmers, but when plants became multicellular they specialized in being sedentary collectors of solar energy. Our animal forebears, on the other hand, remained ambulatory, the better to eat the plants and each other. While plants are enormously successful—the bulk of the earth's biosphere is vegetation, and the largest, most numerous, and longest-lived organisms are plants—they show very little evolutionary tendency toward anything we would recognize as intelligence.

Arthropods—including insects, spiders, and other multilegged crawlers—are certainly mobile. They have highly developed sense organs and nervous systems and exhibit a complexity of behavior similar to that of today's robots. But their exoskeletal construction seems to have limited their metabolism and size, and their nervous systems never exceed about 1 million neurons, compared with 10 million in the smallest vertebrate brain. The social insects have partially overcome their size limitation by coordinating large groups of individuals that act almost as single animals.

Mollusks are as distantly related to us as are arthropods—our common ancestor existed about a billion years ago. They are especially interesting because of greatly contrasting lifestyles among their species. Most mollusks are slow-moving shellfish with filter-feeding lifestyles, but cephalopods—octopus, squid, cuttlefish, and nautilus—swim freely. Octopus and squid have given up their shells to become the most highly mobile invertebrates, with the largest nervous systems. They have imaging eyes like vertebrates, but the eye is a half-sphere attached to flexible body tissue, and the photocells of the retina face toward the lens, rather than away, as in vertebrates. The brain has eight ganglia, arranged in a ring around the esophagus. Octopus and their relatives are swimming light shows, their surfaces covered by a million individually controlled color-changing cells. Small octopus can learn to solve problems like how to open a container of food. Giant squid, as large as whales and with large

nervous systems, are known to exist from a few bloated corpses that have been found floating at sea. But they live permanently at depths beyond present human observation.

Enchinoderms, including starfish, sea urchins, and other seabottom dwellers with five-sided symmetry, have made a stable living for themselves by hunting sessile animals like shellfish and corals. Though closely related to the vertebrates, enchinoderms are much slower-moving and have simpler sense organs and smaller nervous systems. A starfish—in permanent low-gear—hunting a clam looks asleep to human eyes but in time-lapse photography can be seen to stalk its prey, pounce, pry it open, and then begin digesting by everting its stomach into the open shell!!

Fish are very mobile, and have brains larger than most of the invertebrates, but smaller than other vertebrates, possibly because open water is a simple habitat. Reptiles evolved on land and have larger nervous systems. Amphibians are intermediate between fish and reptiles in nervous-system size, as in many other features. Cold-blooded animals like these have a slow metabolism that cannot support nervous systems as large as those of birds or mammals, but for the same reason they can survive with a less frantic lifestyle.

Birds are related to us through an early reptile that lived about 300 million years ago. Their size is limited by the demands of flight, but they have a very energetic metabolism. Though differently structured, they have brains as large as mammals of equal size, and they have comparable learning and problem-solving abilities.

Our last common ancestor with the whales was a primitive rat-like mammal alive 100 million years ago. Some dolphin species have body and brain masses identical to ours, are as good as us at many kinds of problem solving, and can grasp and communicate complex ideas. Sperm whales have the world's largest brains. The edge humans have over other large-brained animals such as elephants and whales may depend less on our individual intelligence than on how effectively that intelligence is coupled to our rapidly evolving, immensely powerful, tool-using culture.

palpable in the robotics lab, and I am confident that this bottom-up route to artificial intelligence will one day meet the traditional top-down route more than half way, ready to provide the real-world competence and the commonsense knowledge that has been so frustratingly elusive in reasoning programs. Fully intelligent machines will result when the metaphorical golden spike is driven uniting the two efforts. A reasoning program backed by a robotics world model will be able to visualize the steps in its plan, to distinguish reasonable situations from absurd ones, and to intuit some solutions by observing them happen in its model, just as humans do. Later I will explain why I expect to see this union in about forty years. For the moment, let us explore some of the artificial fauna at the bottom.

As we have seen, Shakey was not one of them. This robot was an expression of the top-down effort: its specialty was reasoning, while its rudimentary vision and motion software worked only in starkly simple surroundings. At about the same time, though, on a much lower budget, a mobile robot that was to specialize in seeing and moving in natural settings was born at Stanford University's Artificial Intelligence Project, located about eight miles away from Shakey's residence at SRI. John McCarthy founded the Project in 1963 with the then-plausible goal of building a fully intelligent machine in a decade. (The Project was renamed the Stanford Artificial Intelligence Laboratory, or SAIL, as the decade drew nigh and plausibility of the Project drifted away.) Reflecting the priorities of early artificial intelligence research, McCarthy worked on reasoning and delegated to others the design of ears, eyes, and hands for the anticipated artificial mind. SAIL's hand–eye group soon overtook the MIT robotics group and was seminal in the later boom in smart robot arms for industrial uses. A modest investment in mobility was added when Les Earnest, SAIL's technically astute chief administrator, learned of a vehicle abandoned by Stanford's mechanical engineering department after a short stint as a simulated remote-controlled lunar rover. At SAIL it became the Stanford Cart, the first mobile robot controlled by a large computer that did *not* reason, and the first testbed for computer vision in the cluttered, haphazardly illuminated world most animals inhabit. The progeny of two PhD theses (one of them my own), the Stanford Cart slowly navigated raw indoor and outdoor spaces guided by TV images processed by programs quite different from those in Shakey's world.

In the mid-1970s NASA began planning for a robot mission to Mars, to follow the successful Viking landings. Scheduled for launch in 1984, it was to include two vehicles that would rove the Martian surface. Mars is so far away, even by radio, that simple remote control would be either very slow or very risky; the delay between sending a command and seeing its consequence can be as long as forty minutes. If the robot could travel safely on its own much of the time, it would be able to cover much more terrain. Toward this end, Caltech's Jet Propulsion Laboratory (JPL), designer of most of NASA's robot spacecraft, which until then used quite safe and simple automation, initiated an intelligent robotics project. Pulling together methods, hardware, and people from university robotics programs, it built a large, wheeled test platform called the Robotics Research Vehicle, or RRV, a contraption that carried cameras, a laser rangefinder, a robot arm, and a full electronics rack, all connected by a long cable to a big computer. By 1977 it could struggle through short stretches of a rock-littered parking lot to pick up a certain rock and rotate it for the cameras. But in 1978 the project was halted when the Mars 1984 mission was canceled and removed from NASA's budget. (Of course, Mars has not gone away, and JPL is considering a visit there at the end of the millennium.)

Along with the Office of Naval Research, the first and steadiest supporter of artificial intelligence research (and a major reason all the early advances in the field happened in the United States) is the Department of Defense's Advanced Research Project Agency (DARPA). Founded in 1958 after the national humiliation caused by Sputnik, its purpose was to fund far-out projects as insurance against unwelcome technological surprises. In 1981 managers in DARPA decided that robot navigation was sufficiently advanced to warrant a major effort to develop autonomous vehicles able to travel large distances overland without a human operator, perhaps into war zones or other hazardous areas. The number of mobile robot projects jumped dizzyingly, in universities and at defense contractors, as funding for this project materialized. Even now, several new truck-size robots are negotiating test roads around the country—and the dust is still settling.

On a more workaday level, it is not a trivial matter that fixed robot arms in factories must have their tasks delivered to them. An assembly-line conveyor belt is one solution, but managers of increasingly automated factories in the late 1970s and early 1980s found belts, whose routes are difficult to change, too restrictive. Their robots could be rapidly reprogrammed for different jobs, but the material flow routes could not. Several large companies worldwide dealt with the problem by building what they called Automatically Guided Vehicles (AGVs) that navigated by sensing signals transmitted by wires buried along their route. Looking like forklifts or large bumper cars, they can be programmed to travel from place to place and be loaded and unloaded by robot arms. Some recent variants carry their own robotic arms. Burying the route wires in concrete factory floors is expensive, and alternative methods of navigation are being sought. As with robot arms, the academic and industrial efforts to develop mobile robots have merged, and a mind-boggling number of directions and ideas are being energetically pursued.

A Robot for the Masses

The history presented so far is highly sanitized and describes only a few major actors in the new field of robotics. The reality is a witch's brew of approaches, motivations, and, as yet, unconnected problems. The practitioners are large and small groups of electrical, mechanical, optical, and all other kinds of engineers, physicists, mathematicians, biologists, chemists, medical technologists, computer scientists, artists, and inventors, all around the world. Computer scientists and biologists are collaborating on the development of machines that see. Physicists and mathematicians are working to improve sonar and other senses. Mechanical engineers have built machines that walk on legs, and others that grasp with robot hands of nearly human dexterity. Yet all of these fledgling efforts have suffered from poor communication among the various groups, which have not been able to agree upon even a general outline for the field of robotics. Despite the chaos, I expect to see the first mass offering from the cauldron served in time for the new millennium, in the form of a general-purpose robot for the factory—and the home.

In industrialized nations, agriculture and manufacturing is increasingly the province of machines, leaving people free to provide human services for one another. Food and goods have become plentiful and cheap under this arrangement, but many services have increased in cost. Domestic service, once common, is scarce and expensive. Domestic machines such as food processors, vacuum cleaners, and microwave ovens do not fill the void in families where all the adults work outside the home. The need has existed for many decades: *When will there be a robot to help around the house?*

For many years I believed that robot servants, ubiquitous in science fiction, were unlikely in the near future. Households are complex environments with limited resources. The economic return from a mechanical domestic helper would only be a fraction of the value of a robot in a typical industrial role, so the home robot must sell (or rent) for much less. Worse yet, safe and effective operation in the often chaotic home environment is a lot more difficult than in controllable factory settings. Existing robots offer mostly blind, repetitive, potentially lethal motions at a price comparable to that of an entire residence. This enormous gap in price and performance is real, yet I now expect to see a general-purpose robot usable in the home within ten years. The change in my attitude comes partly from research developments of the last few years and partly from a new appreciation of the implications of the concept "general purpose."

Today's industrial robots are more flexible than the fixed automation they sometimes displace, but they do so very few things well that the term "general purpose" hardly applies. Indeed, individual robots are usually bolted to a fixed station, equipped with grippers and sometimes sensors specialized for a certain task, which they henceforth execute, again and again, perhaps for the rest of their existence. The narrowness of their repertoire, besides being boring, greatly limits the number that can be sold. There are less than 100 thousand robots (other than toys) of all makes in the world today. Compare this with 100 million cars, 500 million television sets, or 20 million computers. So few units sold can support only a limited amount of engineering thought and development. The result is a less-than-optimal design at a high price. But not forever. As the number of units produced grows, so does the opportunity and incentive to improve the design of the robots and the details of their production. The costs drop, and the robots become better, incidentally expanding the market and increasing the number of units sold, leading to further improvements. The graph of declining unit cost versus number of units produced is called the *manufacturer's learning curve.*

The potential market for robots will expand enormously when a certain level of general usefulness is achieved. Up to this breakeven point, specialization—the exploitation of the unique circumstances of a job to achieve acceptable performance with minimum complexity—will be the robotic norm. Beyond the breakeven point, the potential market will be large enough that higher profits will go toward more standard designs sold in ever larger numbers. The cheap, mass-produced, high-utility robot will have arrived. We have accumulated enough experience to specify some of the characteristics of this Model T of robots. It will not be intelligent, and it will not come pre-programmed to do many useful tasks. It *will* come from the factory with a sufficient set of mechanical, sensory, and control capabilities that can be conveniently invoked by software specially written for particular applications.

The first major market for such a machine will be in factories, where it will be somewhat cheaper and considerably more versatile than the older generation of robots it replaces. Its improved cost–benefit ratio will allow it to be used in a much wider array of jobs and thus in greater quantities, further lowering its cost. In time it will become cheaper than a small car, putting it within the reach of some households and creating a demand for a huge variety of new software. The robot control programs that actually get various jobs done will come from many different sources, as do programs for today's personal and business computers.

As with personal computers, many successful applications of the general-purpose robot will come as surprises to its makers. We can speculate about the videogame, word-processor, and spreadsheet equivalents of the mass robot era, but the reality will be stranger. To get the guessing game going, consider programs that do light mechanical assembly (from a factory automation company), clean bathrooms (from a small firm founded by former cleaning staff), assemble and cook gourmet meals from fresh ingredients (a collaboration of a computer type and a Paris chef), do tuneups on a certain year of Saturn cars (from the General Motors Saturn service department), hook patterned rugs (by a Massachusetts high school student), weed a lawn one weed at a time, participate in robot races (against other software—programs are assigned a certain physical robot chassis by lot just before the race begins), do detailed earthmoving and stonework (by an upstart construction company), investigate bomb threats (sold to police departments worldwide), deliver to and fetch from a warehoused inventory, help to assemble and test other robots (in several independent stages), and much more. Some of the applications will require optional hardware attachments for the robot, special tools and sensors (such as chemical sniffers), protective coverings, and so on.

It may be that writing applications programs for successive genera-tions of general-purpose robots will become the major human occupa-tion in the early decades of the next century. The skilled plumber, for instance, will be faced with the choice of applying his or her plumbing skills to meet the needs of a few hundred clients or encoding those skills into robot programs that might be sold successfully to thousands or even millions of customers. The first alternative will become increasingly less attractive as a source of income as manual work competes with an ever-increasing number of ever-more-sophisticated robots controlled by ever-better software. The latter course has its risks also—a program may flop in the marketplace, just as inventions, books, music, art, and computer software flop today. On the other hand, a successful program might generate years of income for its author.

Almost everyone has, or can develop, many skills, and each skill can be a potential source of royalties when encoded as a program. Many competing versions of each skill will be marketed and purchased on the basis of utility, cost, personal taste, fashion, and advertising. Each program will have a limited lifetime, destined to be eclipsed by replacements that are either simply better or are designed to operate a more sophisticated new generation of robot. A large secondary industry will spring up to help in the programming process.

Before long programs will be created that make general-purpose robots good learners, teachable, for instance, by leading them through the motions of the task, or by example. The accumulating library of such programs will eventually be a motherlode of encoded hu-man, nonverbal knowledge which can be tapped by the waves of increasingly autonomous robots that will follow the breakeven gener-ation. The expert systems industry has already begun encoding verbal knowledge in this fashion.

Breakeven Locomotion

To be successful, a mass-produced general-purpose robot will require a minimum level of functionality. The first robots of this class need not be able to do everything, or even do most things, very well. They must do enough things well enough to create an open-ended market for themselves, with each drop in price bringing a more than proportional increase in the number of economical applications and units in demand.

Even in highly urban settings, regions of flat, hard ground form an archipelago in a sea of terrain that is variously rough, soft, stepped, or totally impassable. A machine unable to navigate this sea will be trapped on a single island, its potential uses enormously restricted. Our breakeven criterion thus calls for a drive system more capable than standard wheels. Robots with legs are just now showing signs of practicality. The most convincing demonstration to date is by a California company called Odetics, whose six-legged, spiderlike, electrically driven robot can climb out of its truck, onto a hatbox, squeeze through a door, then show off by lifting one end of the truck and dragging it around. This and other promising demonstrations make it likely that practical legged locomotion will be available within a decade.

Legs are a powerful mechanism for movement, but their start–stop nature limits speed and energy efficiency. The Odetics machine, for instance, drains its batteries in under an hour of slow walking. The serious power constraints in a self-contained robot may demand a more frugal drive system. On flat ground wheels are best, offering close to 100% efficiency over a wide range of speed. A compromise solution may be slow legs terminating in wheeled feet—like powered roller skates. The robot would roll on these wheels most of the time but lift its feet over obstacles and up stairs. On rough ground it might plod along slowly in a full walk.

Hitachi, the Japanese electronics giant, experimented in the early 1980s with a particularly simple version of the wheel–foot idea. For use in nuclear reactors, the Hitachi system has five simple "legs," each a vertical, motorized post that telescopes up and down out of the body. The legs are arranged uniformly around the robot, in a regular pentagon. Each ends in a wheel able to steer and drive. Five legs are the minimum that allow a robot to stand stably with any one leg raised, without shifting its weight. The Hitachi machines climb stairs by rolling up to them on five wheels, raising the leading one to the height of the first stair, driving forward until the raised leg is securely over the step, lowering it slightly until the contact is firm, and then continuing with the next nearest leg. On narrow stairs the robot may have its wheels resting on up to three successive steps at the same

Walking Machine

The Odetics "Odex" can walk, climb, squeeze through doorways, or spread for stability. But its power consumption limits it to an hour of mobility per battery charge.

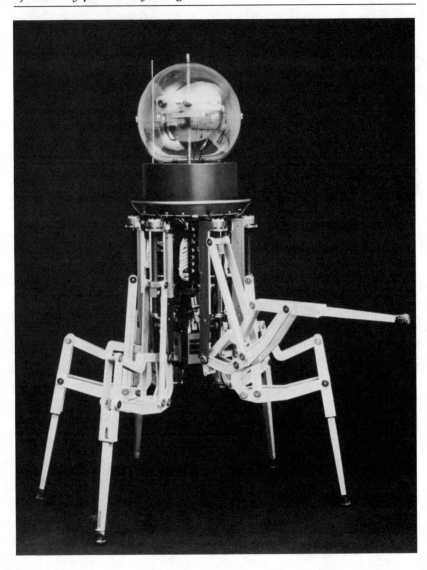

time. A similar procedure carries it over obstacles. The robot can traverse rough ground slowly with the individual wheels riding up and down over the surface irregularities—an active suspension. The upper body of the robot remains perfectly horizontal under normal operation. Time and further research will tell which configuration proves best for the first all-doing robot.

Five Legs

This design for mobility from Hitachi, five steerable wheels on telescoping legs, has more limitations than one with fully articulated legs, but it gets much better mileage on flat surfaces.

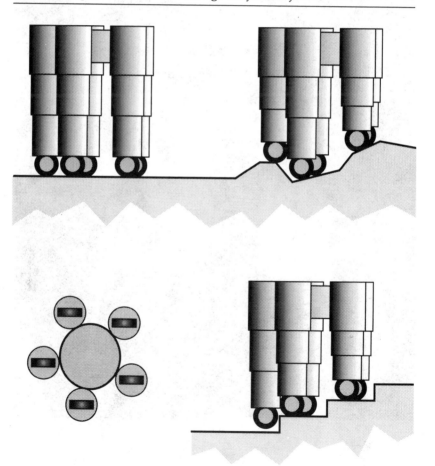

Breakeven Manipulation

Few useful jobs can be accomplished if the robot simply moves about. Productive work calls for the holding and transporting of ingredients, parts, tools, and other things. Industrial manipulators, the most numerous and successful robots to date, have arms that can reach where needed by using about six rotary or sliding joints. If we neglect fine points of weight, power, and control, some of the smaller designs are nearly adequate for the reach needed in the breakeven robot. Since many jobs require bringing pairs of objects into contact, our robot will probably come with at least two arms. A third arm would be advantageous for jobs where the contacting objects must be operated on in some way (electronic hobbyists will recognize soldering as one such job).

Robot hands are not as well developed as robot arms. The industrial manipulators manage to grasp and hold with special fixtures for particular objects or with a simple two-fingered kind of hand called a *parallel-jaw gripper*. Such grippers are easy to operate but can safely grasp only some kinds of rugged object. They are incapable of controlling or changing the orientation of something being held. Our universal robot needs more flexibility.

A few research projects have investigated multifingered grippers that exhibit much greater dexterity. One of the best comes from a ten-year effort by Ken Salisbury, now at MIT. Salisbury's three-fingered robot hand can hold and orient bolts and eggs and manipulate string in a humanlike fashion. He determined the basic configuration and dimensions of the hand with a computer search over different linkages, looking for the minimal set that allowed fingertips to converge on and securely hold arbitrarily shaped small objects. The result has three symmetrically placed fingers that bend much like those of humans. However, because the fingers can bend outward as well as inward, the hand can grip hollow objects from the inside as well as from the outside. The driving forces come through thin steel cables pulled on by a bank of motors some distance down the robot's wrist.

To accomplish feats of even moderate dexterity the hands must "feel" grasped objects. Salisbury is developing hemispherical fingertips for the hand that, through carefully placed internal strain gauges, can sense the magnitude and direction of external forces. The computer programs to plan and to carry out arm and hand motions

Three Fingers

The "Salisbury Hand," a minimal solution for general robot dexterity. Each finger is controlled by three motors. The hand can grip from the outside or, with equal facility, bend its fingers outward to grip a hollow object by its interior.

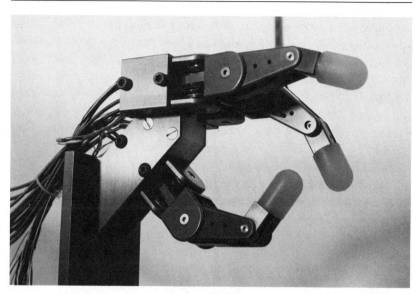

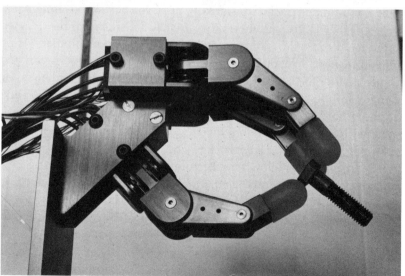

for complex manipulators are still in a tender state. Several programs exist that can plan collision-free arm movements between two points in known clutter. These programs consider a space (the so-called *configuration space*) that describes all possible postures of the manipulator. Each joint adds one dimension to this space, so a manipulator as complicated as Salisbury's has a complex configuration space. A complex space translates into an expensive and time-consuming search for a good path. Running times of minutes to hours are typical, but better algorithms continue to be found, and computers continue to become faster.

Breakeven Navigation

The mechanical ability to move is only part of the problem of mobility. One must also be able to find and return to specific locations and avoid dangers in transit. I have been working on this issue for most of my career and am happy to report that some good solutions are developing. My thesis work at Stanford during the 1970s was on programs that were intended to let the Stanford Cart find its way through cluttered rooms and outdoor spaces. The first version of such a program, in 1976, obtained its view of the world through one TV camera on the robot. By locating distinctive areas in the TV image and tracking them as the robot moved, the program was able to estimate their distance and the extent of its own motion. It constructed a sparse three-dimensional map of its surroundings, identified obstacles in it, and planned a path to its destination that stayed clear of them. The program then moved the robot about a meter along that path, looked, mapped, planned, and moved again. In repeated cautious lurches, the Cart was to creep safely to its destination.

Unfortunately, the program didn't work. About one lurch in four, the part of the program that attempted to estimate the robot motion from the changing image made a mistake, misidentified areas being tracked, reported an incorrect robot motion, and messed up the slowly building map. The chance of successfully crossing a large room, a journey of perhaps thirty lurches, was so small as to be negligible. In 1979 I tried again, with a new program aided by a small amount of new hardware, a mechanism that precisely moved the camera from side to side along a track. With it, the driving program was able to obtain several images of the scene without moving the whole robot, much as a human obtains two images, one from each eye. By carefully exploiting the extra information to prune away errors, the program improved the success rate for a single lurch to nearly 100%. The robot was now often able to successfully complete the thirty lurches to cross a room to the desired destination and show a correct map on a display screen. About one time in four, however, it failed, either because the aggressive error-pruning had removed a real obstacle from the map and the robot had collided with it, or because, in spite of the pruning, errors had crept in and confused the robot's idea of its position. Good enough for my thesis, perhaps, but not good enough for a robot to do complex tasks that would, at the minimum, require it to cross rooms many times.

In 1980 I moved to Carnegie Mellon University, to continue the research under the auspices of its new Robotics Institute. Two graduate students, Chuck Thorpe and Larry Matthies, examined and greatly improved the old program, increasing both its speed and its accuracy tenfold. When everything went well, it was now able to report the position of the robot (a new one we call Neptune) to a few centimeters accuracy. Unfortunately, things did not always go well, and the failure rate remained stubbornly unchanged. The robot still crossed the room correctly only about three times out of four.

In 1984 our group agreed to do some research for a new company, Denning Mobile Robotics, Inc., in Massachusetts, that was developing a robot security guard (more accurately, a roving burglar alarm). Instead of a camera, the robot was equipped with a beltlike ring of sonar range sensors like those found in Polaroid cameras. These had already been found to be very useful for detecting the presence and general direction of nearby obstacles, thus allowing the robot to avoid them. Our aim was more ambitious, however. Instead of merely sensing imminent collisions, could the continuously active sonar system be used to build a map of the surroundings that could direct accurate point-to-point navigation, as (three times out of four) our vision-guided programs could do? The sonar units each emit an ultrasonic chirp of sound over a wide cone and report the time to the first echo they hear. This time is proportional to the distance of the nearest object within the cone. The distance to the object may be accurate to better than a centimeter, but since the cone subtends an angle of about 30°, the side-to-side position is still highly uncertain. This is very different from the almost pinpoint measurements possible from TV cameras, and so the program methods developed for the Cart could not be used.

Although a single sonar reading can tell a program only a little about the position of the thing that caused the echo, it maps out a large volume of empty space in front of that thing. When readings from different sensors overlap, the empty region indicated by one

Autonomous Navigation
The Denning Sentry is a commercial product that can patrol a large warehouse or office complex every night for months without human intervention, guided by light-emitting beacons and a sonar image of its surroundings. By day, it recharges itself in a special booth.

reading may restrict the possible location of the echo-causing object indicated by the other. Hundreds or thousands of readings from different positions, taken together, might be able to build detailed maps in spite of the fuzziness of individual sensors. Because the sensitivity of a sonar sensor falls off smoothly from the middle to the edges of its cone, it seemed best to do the mixing with probabilities. Alberto Elfes, another graduate student, and I wrote a program based on these ideas and were astonished when it drove the robot much more reliably than the old TV-guided program. Yet another student, Bruno Serrey, along with Larry Matthies, then found a way to use the probabilistic approach for TV data and again discovered that it worked remarkably better than the old approach.

Our new method represents the space around the robot as a grid of cells, each containing the probability, based on all available sensor readings, that a corresponding cell in space is occupied by matter. A reading may lower the probability of some cells (for instance, those belonging to the interior of a sonar cone) and raise others (such as those on the range surface of the sonar reading). It provides a convenient way to combine the results of different kinds of readings, and indeed Elfes and Matthies recently demonstrated a program that builds maps from both sonar and TV data. Using these new methods, our robots can now travel long distances almost flawlessly. With a new mathematical foundation for this approach, and a new, very fancy, robot, Uranus, to continue the work, I feel extremely confident that navigation will be more than adequately in hand within the ten-year timeframe of the universal robot.

Breakeven Recognition

The sensory system has another vital function: the recognition and localization of specific objects in the robot's surroundings. Recognized objects may be small things destined later to be picked up by one of the hands or large objects that serve as landmarks or work locations. Imagine a process whereby objects are described by shape and surface characteristics and the robot's recognition system looks for one object at a time. A tentative identification can be confirmed by viewing the scene from a different point. The result is a description of the position and orientation of the object suitable for use by the program that controls grasping by the hands.

Object Finding

3DPO (for Three-Dimensional Parts Orientation) is a program that finds particular parts in a clutter of other parts. This sequence of images is: (1) a three-dimensional computer model of the part to be found; (2) a TV picture of a jumble of actual parts; (3) a computer image of the same parts where brightness now indicates how close to the camera is each bit of visible surface; (4) the computer's deduction of the major surface boundaries of the jumble; and (5) the computer's fit of the part model to actual occurrences of the part in the jumble.

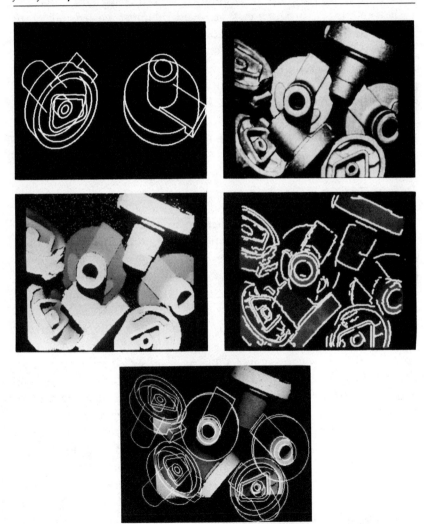

Computer vision is by far the most promising vehicle for this identification ability. The key operation is identifying a specific object in a mass of clutter. Vision research for industrial robots has produced partial solutions to the so-called "bin-picking" problem. Bin-picking programs let a computer identify predefined objects in a jumble in a TV image, even if the objects partially occlude each other, so that they can be removed one at a time by a manipulator. A General Motors research group in the 1970s demonstrated a system that worked if the overlapping parts mostly lay flat. It was too slow and unreliable to be practical in production, but it did demonstrate feasibility. In the last several years many groups in the United States and Japan have unveiled programs that can identify simple objects on the basis of three-dimensional data obtained from a camera looking at a scene illuminated by special devices that generate stripes or grids of light. On contemporary computers, these systems take many minutes to make their less-than-satisfactory identifications. Yet it is likely that the minimal requirements for our robot will be met within our ten-year timeframe.

Processing and Coordination

The best prototypes for the low-level sensory and movement-planning parts of our future robot all consume many minutes of computer time on a good microcomputer. This is partly a measure of the patience of the researchers; processes that run for much longer than an hour are simply too difficult to investigate effectively, while simpler, faster programs are not very interesting because they work less well. On the other hand, the running times do tell us something about the difficulty of the breakeven criteria. A robot that spends up to an hour considering every simple move is clearly unacceptable, but a few seconds would be tolerable. A computer able to do a billion operations per second, with a billion bytes of main memory, would be enough. This is about the power of the largest supercomputers that have been built to date, and a few hundred times faster than the best microcomputers. The continued computer evolution should deliver it in a microcomputer within a decade. Depending on the progress in various lines of development, the power may be spread among few or many individual processing units, and may depend on a significant fraction of specialized hardware, for instance, circuits to do low-level vision processing. The exact hardware configuration is unimportant for our purposes here.

Our work at Carnegie Mellon with integrated tasks for a mobile robot suggests that basic processes should be organized into modules that run concurrently. A navigation program driving the robot to a desired location might, for example, coexist with ones that watch out for surprises and dangers. If a stairwell-detecting module concludes that hazard is near, it would take over control of the robot until the danger was past.

A Sensible Robot

Here is a possible configuration for our mass robot. It moves on five leg-wheels of the Hitachi design and has two arms with Salisbury hands. Topped by a pair of color TV cameras, it has an unobtrusive array of sonar sensors to sense the world in directions not covered by the cameras. It carries an inexpensive laser gyroscope to help with navigation, and it is controlled by a computer system able to do at least a billion operations per second. Integral with the computer hardware is a software operating system that allows multiple simultaneous processes. Built-in programs permit objects in the world to be described, visually identified in or out of clutter, and picked up. A navigational system can be asked to build, store, retrieve, and compare maps of the surroundings and to bring the robot to specific locations.

Readers familiar with personal computers may recognize the similarity to operating-system utility functions, especially the graphic toolbox in the Apple Macintosh. These capabilities in the robot are orchestrated by application software for (one hopes) an astonishing variety of specific jobs; software is supplied by many independent vendors. Again the similarity to personal computers is clear. One might eventually hope for integrated software packages that allow the robot to switch quickly and automatically from one task to another, making it a more autonomous mechanical servant.

The Convergent Evolution
of Emotions and Consciousness

The machines we have been considering behave in a predictable way that we might describe as mechanical or insectlike. Will robots

A General-Purpose Robot

This caricature of a first-generation general-purpose robot shows the major systems: Locomotion with limited stair and rough-ground capability, general manipulation, stereoscopic vision, coarse 360° sensing for obstacle avoidance and navigation. Not shown is the computer hardware and software that will be required to animate this assembly.

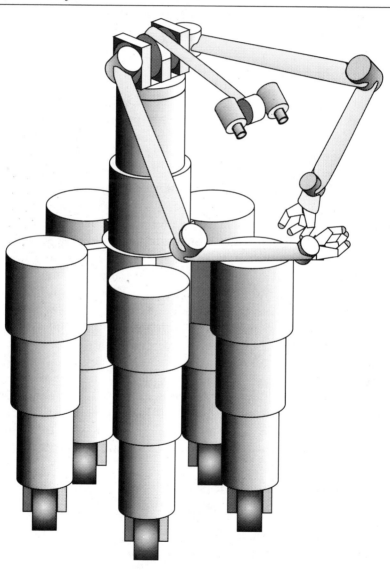

continue to display this predictability as they become more complex, or will they develop something akin to the richer character of higher animals and humans?

As we have seen, the more advanced control programs in today's roving robots use data from sensors to maintain representations, at varying levels of abstraction and precision, of the world around the robot, of the robot's position within that world, and of the robot's internal condition. The programs that plan actions for the robot manipulate these world models to weigh alternative future moves. The world models can also be stored from time to time and examined later, as a basis for learning.

A verbal interface keyed to these programs would meaningfully answer questions like "Where are you?" ("I'm in an area of about twenty square meters, bounded on three sides, and there are three small objects in front of me") and "Why did you do that?" ("I turned right because I didn't think I could fit through the opening on the left.") In our lab, the programs we have developed usually present such information from the robot's world model in the form of pictures on a computer screen—a direct window into the robot's mind. In these internal models of the world I see the beginnings of awareness in the minds of our machines—an awareness I believe will evolve into consciousness comparable with that of humans.

The term *convergent evolution* is used by evolutionary biologists whenever species that are only very distantly related independently develop similar characteristics, presumably in response to similar environmental pressures. Eyes are an example of convergent evolution; they have evolved over 40 different times in the animal kingdom. What was necessary was the presence of light-sensitive cells and selection pressures favoring the survival of animals that could see, however dimly at first. If a function of the nervous system as complex as vision can evolve so many different times when environmental pressures are right, what about emotions and consciousness? Unlike vision, these features of the human mind have no incontrovertible external manifestation and indeed lack a precise definition. Their existence in animals, and even in humans, has been questioned by a generation of behavioral psychologists. Yet animal ethologists such as Donald Griffin find the concepts useful in explaining animal behavior. If an animal acts as I do when I am afraid, is it not reasonable to call its mental state "fear"? If it chooses from among several complex alternatives in dealing with a novel situation, as I would consciously weigh my options in the same circumstance, why not ascribe "consciousness" instead of some other mechanism with a different name but the same effect?

Consider the following thought experiment. Suppose we wish to make a robot that can execute some general task such as "Go down the hall to the third door, go in, look for a cup, and bring it back." Our most pressing need would be a computer language in which to specify complex tasks for the rover and a hardware and software system to embody it. Sequential control languages successfully used with industrial manipulators might seem to be a good starting point. But paper attempts to define the structures and primitive actions required for mobility would reveal that the linear control structure of these state-of-the-art languages, though adequate for a robot arm, would prove to be inadequate for a rover. The essential difference is that a rover, in its wanderings, is regularly "surprised" by events which it cannot anticipate but with which it must deal. This requires that contingency routines be activated in arbitrary order and run concurrently, each with its own access to the needed sensors, effectors, and the internal state of the machine, and a way of arbitrating their differences. As conditions change, the priority of the modules changes, and control may be passed from one to another.

A request to our future robot to go down the hall to the third door, go in, look for a cup, and bring it back might be implemented as a module, FETCH-CUP, that looks very much like a program written for the arm-control languages (which in turn look very much like programming languages such as Algol or Basic), except that another module, COUNT-DOORS, would run concurrently with the main routine. Consider the following outline for such a program.

Module COUNT-DOORS:

  Check the robot's surroundings for doors

  Add one to the variable DOOR-NUMBER each time a new door is located

  Record the location of the new door in the variable DOOR-LOCATION

Module GO-FETCH-CUP:

  Step 1: Record the current location of the robot in the variable START-LOCATION

  Step 2: Set the variable DOOR-NUMBER to zero

Step 3: Wake up the COUNT-DOORS module

Step 4: Drive the robot parallel to the right-hand wall until DOOR-NUMBER is three or greater

Step 5: Cause the robot to face the location in the variable DOOR-LOCATION

Step 6: If the robot is facing an open door, go to Step 10

Step 7: If the robot is not facing a door, subtract one from DOOR-NUMBER and go to Step 4

Step 8: If the robot is facing a closed door, try to open it

Step 9: If the door fails to open, say "knock knock" and go to Step 6

Step 10: Drive the robot through the open door

Step 11: Check the robot's surroundings for cups; if there are none, go to Step 15

Step 12: Record the location of the nearest cup in CUP-LOCATION

Step 13: Drive the robot to within reach of the CUP-LOCATION

Step 14: Pick up the cup at CUP-LOCATION; if this fails go to Step 15

Step 15: Go back and face the door at DOOR-LOCATION

Step 16: If the robot is facing a closed door, try to open it

Step 17: If the door fails to open, say "knock knock" and go to Step 16

Step 18: Drive the robot through the open door

Step 19: Return to START-LOCATION

Step 20: Put the robot to sleep

So far so good. We activate our program, and the robot obediently begins to trundle down the hall, counting doors. It correctly recognizes the first one. The second door, unfortunately, is decorated with garish posters, and the lighting in that part of the corridor is poor, so our experimental door-recognizer fails to detect it. The wall-follower, however, continues to operate properly and the robot continues on down the hall, its door count short by one. It recognizes door 3, the one we had asked it to go through, but thinks it is only the second, so continues. The next door is recognized correctly and is open. The program, thinking it is the third one, faces it, and proceeds to go through. This fourth door, sadly, leads to the stairwell, and the poor robot, unequipped to travel on stairs, is in mortal danger.

Fortunately, there is another module in our concurrent programming system called DETECT-CLIFF. This program is always running and checks ground position data incidentally produced by the vision processes and also requests sonar and infrared proximity checks on the ground. It combines these, perhaps with an *a priori* expectation of finding a cliff set high when operating in dangerous areas, to produce a number that indicates the likelihood there is a drop-off in the neighborhood. A companion process DEAL-WITH-CLIFF, also running continuously but with low priority, regularly checks this number and adjusts its own priority on the basis of it. When the cliff probability variable (perhaps we'll call it VERTIGO) becomes high enough, the priority of DEAL-WITH-CLIFF will exceed the priority of the current process in control, GO-FETCH-CUP in our example, and DEAL-WITH-CLIFF takes over control of the robot. A properly written DEAL-WITH-CLIFF will then proceed to stop or greatly slow down the movement of the robot, to increase the frequency of sensor measurements of the cliff, and to back away slowly from it when it has been reliably identified and located.

Now there is a curious thing about this sequence of actions. A person seeing them, not knowing about the internal mechanisms of the robot, might offer the interpretation, "First the robot was determined to go through the door, but then it noticed the stairs and became so frightened and preoccupied it forgot all about what it had been doing." Knowing what we do about what really happened in the robot, we might be tempted to chastise this poor person for using such sloppy anthropomorphic concepts as determination, fear, preoccupation, and forgetfulness in describing the actions of a machine. We could chastise the person, but in my opinion that would be wrong. The robot came by its foibles and reactions as honestly as any living animal; the observed behavior is the correct course of action for a being operating with uncertain data in a dangerous world. An octopus in pursuit of a meal can be diverted by subtle threats of danger in just the way the robot was. The invertebrate octopus also happens to have a nervous system that evolved entirely independently of our own vertebrate version. Yet most of us feel no qualms about ascribing qualities like passion, pleasure, fear, and pain to the actions of the animal. I believe that we have in the behavior of a person, an octopus, and a robot a case of convergent evolution. The needs of the mobile way of life have conspired in all three instances to create an entity that has modes of operation for different circumstances and that changes quickly from mode to mode on the basis of uncertain and noisy data prone to misinterpretation. As the complexity of mobile robots increases, their similarity to animals and humans will become even greater.

Hold on a minute, you say. There may be some resemblance between the robot's reaction to a dangerous situation and an animal's, but surely there are differences. Isn't the robot more like a startled spider, or even a bacterium, than like a frightened human being? Wouldn't it react over and over again in exactly the same way, even if the situation turned out not to be dangerous? You've caught me. I think the spider's nervous system is an excellent match for robot programs possible today. (We passed the bacterial stage in the 1950s with light-seeking electronic turtles.) This does not mean that concepts like thinking and consciousness must be ruled out, however.

In his book *Animal Thinking*, Griffin reviews evidence that much animal behavior, including the behavior of insects, can be explained economically in terms of consciousness: an internal model of the self, surroundings, and other individuals that, however crudely, allows consideration of alternative actions. For instance, bees, as Otto von Frisch discovered, communicate direction, distance, and desirability of a food source to other members of a hive by the direction, length, and vigor of each burst in a "waggle dance." Martin Lindauer extended Frisch's observations to cases where a swarm from an overpopulated colony seeks out a new site. A worker from the swarm flies out in search of suitable cavities and returns when it has found and meticulously explored one. It then performs a waggle dance, on the surface of the swarm, describing the location and suitability of its discovery. Meanwhile, other workers tell of other locations. Promising sites are visited and carefully examined by other members of the swarm, who return to tell the tale. A worker telling of one site is unaffected by another bee sending the same message but can be "converted" by a sufficiently emphatic and repeated display describing a different location. The debate rages for several days, with repeated visits to a dwindling number of candidate sites, until near unanimity is reached. The entire swarm then flies to take up residence in the winning cavity. This performance might be explained if we postulate a simple map in the brain of each bee describing locations and their desirability, maps which can be modified by the complex experiences of exploration or the simpler ones of communication and which can become the basis of choice.

An internal model of the world complex enough to allow choices in behavior—whether or not we call this model "consciousness"— is what roboticists are currently trying to achieve in their roving robots. In fact, robotics research is too practical to seriously set itself the explicit goal of producing machines with such nebulous and controversial characteristics as emotion and consciousness. It would be enough if our machines could make a living in the face of the many surprises, setbacks, opportunities, barriers, and competitors they will encounter in the world. But natural selection, the guiding mechanism of Darwinian evolution, is equally utilitarian, and yet here we are, with feelings and a sense of self.

In *The Growth of Biological Thought*, the evolutionary biologist Ernst Mayr points out that both living and nonliving systems "almost always have the property that the characteristics of the whole cannot (even in theory) be deduced from the most complete knowledge of the components, taken separately or in other partial combinations." *Emergence*—this appearance of novel properties in whole systems— has often been invoked to explain such difficult biological realities as mind, consciousness, and even life itself. Here is how I imagine some of the more mysterious mental experiences that we associate with human beings might emerge in our machines as we pursue utilitarian functionality.

Learning

When tickled, the sea slug Aplysia withdraws its delicate gills into its body. If the tickling is repeated often, with no ill effect, Aplysia gradually learns to ignore the nuisance, and the gills remain deployed. If, later, tickles are followed by harsh stimuli, such as contact with a strong acid, the withdrawal reflex returns with a vengeance. Either way, the modified behavior is remembered for hours. Aplysia has been studied so thoroughly in the last few decades that the neurons involved in the reflex are well known, and the learning has recently been traced to chemical changes in single synapses on these neurons. Larger networks of neurons can adapt in more elaborate ways, for instance by learning to associate specific pairs of stimuli with one an- other. Such mechanisms tune a nervous system to the body it inhabits and to its environment. Vertebrates owe much of their behavioral flexibility to an elaboration of this arrangement, systems that can be activated from many locations that encourage and discourage future repetitions of recent behaviors. Though the neural architecture of these systems in vertebrates is not understood, their effect is evident in the subjective sensations we call pleasure and pain.

Existing robot systems are, at best, configured to learn a few specific things from their environment—a simple sequence of moves, the location of an expected component, the position of nearby obstacles, sometimes a few parameters for best controlling a motor or interpreting a sensor. There is little point in having them learn to orchestrate their actions in complicated ways when we can hardly program them to do one thing at a time well. Yet this primitive state of affairs will not last forever. Beginning, perhaps, with the universal robot I described earlier, it will become desirable to add some very general learning abilities.

A robot's safety and usefulness in a home would be greatly enhanced if it could learn to avoid idiosyncratic dangers and exploit opportunities. If a particular door on a certain route is often locked, it might be worthwhile if the robot could learn to favor a longer but more reliable path. A job might be done more effectively if the changing location of a needed ingredient could be learned or even anticipated from subtle clues. It is impossible to explicitly program the robot for every such eventuality, but much could be accomplished by a unified conditioning mechanism which increased the probability of decisions that had proven effective in the past under similar circumstances and decreased it for ones that had been followed by wasted activity or danger.

The conditioning software I have in mind would receive two kinds of messages from anywhere within the robot, one telling of success, the other of trouble. Some—for instance indications of full batteries, or imminent collisions—would be generated by the robot's basic operating system. Others, more specific to accomplishing particular tasks, could be initiated by applications programs for those tasks. I'm going to call the success messages "pleasure" and the danger messages "pain." Pain would tend to interrupt the activity in progress, while pleasure would increase its probability of continuing.

The messages also would provide input to a program that used statistical techniques to compactly "catalog" the time, position, activity, surroundings, and other properties known to the robot that preceded the signal. A "recognizer" would constantly monitor these variables and compare them with entries in the catalog. Whenever a set of conditions occurred that was similar to those that had often preceded pain (or pleasure) in the past, the recognizer would itself issue a somewhat weaker pain (or pleasure) message. In the case of pain, this warning message might prevent the activity that had caused trouble before. In time the warning messages themselves would accumulate in the catalog, and the robot would begin to avoid the steps that led to the steps that caused the original problem. Eventually a long chain of associations like this could head off trouble at a very early stage. There are pitfalls, of course. If the strength of the secondary warnings does not weaken sufficiently as the chain lengthens, pain could grow into an incapacitating phobia and pleasure into an equally incapacitating addiction.

Besides allowing the robot to adapt opportunistically to its environment, a pleasure–pain mechanism could be exploited by applications programs in more directed ways. Suppose the robot has a spoken word recognizer. A module that simply generates a pleasure signal on hearing the word "good" and a pain message on hearing "bad" would allow a customer to easily modify the robot's behavior. If the robot was making a nuisance of itself by vacuuming while a room was in use, a few utterances of "bad!" might train it to desist until conditions changed, for instance at a different time of day or when the room was empty.

A robot with conditioning software could be programmed to train itself. If a task in an application program required that a certain kind of container be opened, it would be possible to write a detailed list of instructions describing just how to hold, turn, and pull to get the job done. Alternatively, a robot at the factory could be programmed to pick up many such containers, one after another, and randomly push, twist, shake, and pull each one until it either opened or broke. The training program would recognize both situations and issue a pleasure message in one case and a pain signal in the other before going on to the next container. Gradually the conditioning system would inhibit those sequences that caused breakage and facilitate those that were successful. An abstracted version of the training catalog for the session could then be inserted into the final application program in place of explicit instructions, combined, perhaps, with catalogs for other parts of the task developed on other robots.

Infinite patience would be an asset in a training session, but it could be exasperating in a robot in the field. In the cup-fetching program I described earlier, you may have noted that if the robot finds the door closed and is unable to open it, it simply stands there and repeats "knock knock" without letup until someone opens the door for it. A robot that often behaved this way—and many present-day robots do— would do poorly in human company. Interestingly, it is possible to trick insects into such mindless repetition. Some wasps provide food for their hatching eggs by paralyzing caterpillars and depositing them in an underground burrow. The wasp normally digs a burrow and seals its entrance, then leaves to hunt for a caterpillar. Returning with a victim, she drops it outside the burrow, reopens the entrance, and then drags it in. If, however, an experimenter moves the caterpillar a short distance away while the wasp is busy at the opening, she retrieves her prey, and then again goes through the motions of opening the already open burrow. If, while she is doing this, the experimenter again moves the caterpillar away, she repeats the whole performance. This cycle can apparently be repeated indefinitely, until either the wasp or the experimenter drops from exhaustion. A robot could be protected from such a fate by a module that detects repetitious behavior and generates a weak pain signal on each repetition. In the example, the door knocking would gradually become inhibited, freeing the robot for other pending tasks or inactivity. The robot will have acquired the ability to become bored.

Modules that recognize other conditions and send pain or pleasure messages of appropriate strength would endow a robot with a unique character. A large, dangerous industrial robot with a human-presence detector sending a pain signal would become shy of human beings and thus be less likely to cause injury. A module that registered pleasure on encountering new debris, and pain on seeing it subsequently, might enable a cleaning program to become very creative and aggressive in its battle against filth.

Imagery

Fast-learning robots would be able to handle programs that had a great many alternative actions at each stage of a task—such alternatives would give the robot a wide margin for creativity. But a robot with only a simple conditioning system would be a slow learner. Many repetitions would be required to elicit statistically significant correlations in the conditioning catalog. Some situations in the real world are unforgiving of such a leisurely approach. A robot that repeatedly wandered onto a public road, being slow to register the danger of that location, might suddenly be converted into scrap metal. A robot, or software, that was slow in adapting to changing conditions or opportunities in a house could lose the battle for economic survival against a swifter product from another manufacturer.

Learning could be greatly enhanced by the addition of another major module, a general *world simulator*. Now, even the bare-bones universal robot I outlined uses simulation to some extent. To safely reach its destination, a program in the universal robot consults its internal map of the surroundings and considers many alternative paths to find the best. These ponderings are simulations of hypothetical robot actions. Similar processes go on when the robot decides how to pick up an object or when it considers possible interpretations of what it sees with its cameras. But each of these procedures is specialized, models only one aspect of the world, and can be used for only one function. Suppose the robot had a much more powerful simulator that permitted complex hypothetical situations involving the robot and many aspects of its surroundings to be modeled. An application program might use such a simulator to check out a proposed action for safety and efficacy, without endangering the robot.

But things get really interesting when events in the simulator are fed to the conditioning mechanism. Then, a disaster in the simulator (for instance a simulated tumble of the robot) would *in real life* condition the robot to avoid the simulated precursor event (let's say loitering at the head of a simulated stairwell). The robot could thus prepare for many future problems and opportunities by simulating possible scenarios in its idle time. Such scenarios might be simply variations on the day's real events. So equipped, the robot will have the capacity to remember, to imagine, and to dream.

Imagination via simulator is useful only if the simulator makes reasonably accurate predictions about the real world. Doing so requires much knowledge about the world, and I imagine that the competitive development of increasingly good simulators will be a major part of the research effort of the early twenty-first-century robotics industry. Robot companies will observe the foibles of their robots in the laboratory and the field, and tinker with the simulators, the better to model those aspects of the world important for robot performance. Evolution did the same for us in the eons of our development. The simulators will come from the factory loaded with generic knowledge, but they will also be required to learn the idiosyncrasies of each new location. Advanced robots may find themselves working with other robots and with people. Such interaction could be made more effective if the simulators on these machines could predict the behavior of others to some extent. Part of the prediction might involve roughly modeling the other's mental state, so that its reactions to alternative acts could be anticipated. A rich new arena opens up once there is an internal model of another being's state of mind. For instance, a module that generated pain messages when it detected distress in a mental model in the simulator would condition the robot to act in a kindly manner. And a robot might find itself admonished for inappropriately ascribing "robotomorphic" feelings and intentions to other machines, or to humans!

It would, of course, be as easy to program robots to commit crimes as to perform socially sanctioned tasks, and legal ways of assigning blame when this happens will no doubt be devised. But complex robots will sometimes get into trouble on their own initiative. Imagine a simulator-equipped robot that has several times in the past suffered dire consequences from being unable to recharge its batteries in time. It will thus be especially strongly conditioned against allowing its power to run low. Suppose it finds itself locked out of its owners' home, its battery charge fading. The robot's simulator will churn through different scenarios furiously, searching for a solution—a combination of actions that will result in a recharge. As combinations of conventional behaviors fail to get it closer to its goal, the simulator search expands to more unusual possibilities. The neighbors' house is nearby, its door may be open, and there will be power outlets inside— the simulator discovers a scenario that takes the robot to those outlets. There is pain associated with leaving home territory, and with the trouble it may cause, but it is more than balanced by the pleasure and great release from pain in the possibility of finding a recharge. The robot repeatedly runs a simulation of the trespass of the neighbors' house, each time strengthening its conditioning for the steps involved, making the act itself increasingly likely. Eventually the conditioning is sufficient, and the robot begins on a course that is likely to lead it into more trouble than its imperfect simulator anticipated. It will not be the first creature to have been driven to a desperate act by a great need.

We could carry this speculative evolution further, gradually endowing our feeling robots with intellectual capabilities similar to those of humans. I expect, however, that by the time the robots are ready for them, superb intellectual capabilities will be available for wholesale purchase from the traditional artificial-intelligence industry, which will have been pursuing its top-down strategy in parallel with the bottom-up evolution of the robots. The marriage may take many years to consummate fully, raising issues such as how the reasoning system can best access the simulator to derive flashes of intuition, and how reasoning should influence the conditioning system so as to be able to override the robot's instincts in exceptional circumstances. The combination will create beings that in some ways resemble us, but in other ways are like nothing the world has seen before.

2 | *Powering Up*

DURING the 1970s, while I was a graduate student, it seemed to me that the processing power available to artificial intelligence programs was not increasing very rapidly. In 1970 my work was done on a Digital Equipment Corporation PDP-10 mainframe computer serving a community of perhaps thirty people. By 1980 my computer was a DEC KL-10, five times as fast and with five times the memory of the old machine but serving twice as many users. Worse, the little remaining speedup seemed to have been absorbed in computationally expensive convenience features: fancier time sharing and high-level languages, graphics, screen editors, mail systems, computer networking, and other luxuries that had become necessities.

Several effects together produced this state of affairs in computing hardware. Support for university science in general had wound down in the aftermath of the Apollo moon landings and the Vietnam war, leaving the universities to limp along with aging equipment. The same conditions caused a recession in the technical industries: unemployed engineers opened fast-food restaurants instead of designing computers. The initially successful problem-solving thrust in artificial intelligence had not yet run its course, and it still seemed to many that existing machines were powerful enough—if only the right programs could be found. Yet progress in the research itself became slow, difficult, and frustrating, and many of the best programmers were drawn into the more rewarding activity of building attractive, but computationally expensive, software tools, whose success spawned yet more tool building.

If the 1970s were the doldrums for computing hardware, the 1980s have more than compensated. Just as artificial intelligence was given its first boost in the 1960s by the Russian leap into space, the second

51

stage was ignited in the present decade by the Japanese leap into the American marketplace. The Japanese industrial successes focused attention worldwide on the importance of technology, particularly computers and automation, in modern economies. American industries and government responded with research dollars. The Japanese stoked the fires, under the influence of a small group of senior researchers, by boldly announcing a major initiative toward future computers, the so-called Fifth Generation project, which would expand in the most promising American and European research directions. The Americans responded with more money.

Besides this economic boon, integrated circuitry had evolved far enough by the 1980s that an entire computer could fit on a chip. Suddenly computers were affordable by individuals, and a new generation of computer customers and manufacturers came into being. On the other end of the scale, supercomputers, once reserved for a handful of government labs and agencies, became fashionable in hundreds of industry and research settings. Across the spectrum of size, the computer industry became lucrative and competitive as never before, with new generations of faster, cheaper machines being introduced at a frenetic rate.

How much further must this evolution proceed until our machines are powerful enough to approximate the human intellect? Too little is known about both the overall functioning of the human brain and how an intelligent computer would operate to make this estimate directly. I have approached the problem indirectly by comparing a fragment of the nervous system that is moderately well understood—the retina of the eye—with computer vision programs that do approximately the same job. I then extrapolate the ratio from that comparison to the whole brain, in order to obtain the computing power required in a machine that would mimic it. The time of arrival of a machine of that power is then estimated by extending into the future the trendline of computer power per unit cost as it has developed during this century.

The computer estimates I will use in making these comparisons are from my own research. The neurobiology is abstracted from John Dowling's authoritative book *The Retina* and Stephen Kuffler and John Nicholls' classic textbook *From Neuron to Brain*. The numbers are precarious because both computer vision and our understanding of biological vision (not to mention other brain functions) are in their infancy. Many fundamentals remain mysteries in this complex domain. Fortunately, my comparison does not require fiendish precision; errors of 100 times either way will make little qualitative difference in relation to the large logarithmic scales of this chapter. Besides, I also hope that some of my errors will be in opposite directions and thus partly cancel. There are some dangerous curves in this joyride to human equivalence, so hold on!

Neural Circuitry

The retina is really an elongated extension of the brain. But its location at the back of the eyeball, some distance from the bulk of the brain, has made it comparatively easy to study, even in living animals. Removed from the body, it can be kept functioning for hours, with its inputs and outputs highly accessible. Transparent and thinner than a sheet of paper, the retina can be stained with dyes to make specific neurons visible to light and electron microscopes. For these reasons, the retina is probably the best-studied piece of the vertebrate nervous system. We will look at it in some detail, but first some background about nerve cells.

All neurons, like other cells, are daunting mechanisms. They begin life by differentiating from stem cells early in the growth of an embryo, then go through repeated cycles of crawling amoeba-like to precise destinations throughout the body and dividing and differentiating further. When they reach their final location, they extend fibrous growths that seek out specific connections with other neurons, through junctions called *synapses*. Different subpopulations of neurons differ radically in geometry, size, and function. Some neurons have thousands of small fibers called *dendrites* and may be host to hundreds of thousands of synapses. One fiber, known as the *axon*, can grow to several centimeters in length, a million times the cell's original size.

A typical neuron receives messages on its dendrites and issues them on its axon, which can branch at its end. It signals by means of electrical potential differences of a few millivolts across its outer membrane. The voltage is maintained in this wet, electrically conductive environment by molecular ion pumps in the membrane that move charged potassium, sodium, calcium, chloride, and other ions in and out of the cell. The pumps are activated or inhibited by small molecules called

The Retina

A cross-sectional view through the 1/2 millimeter thickness of a human retina. The lens is toward the bottom, and light must penetrate three layers of image-processing circuitry to reach the light-sensitive photocells at the top. The horizontal cells compute average intensity over large areas of photocells; the bipolar cells subtract such large-area averages from averages over smaller areas to produce "center-surround" signals that are strongest at object boundaries. Some amacrine cells further enhance the center-surround signals; others signal moving objects. The ganglion cells complete the computations and send the results as pulses along their long axons to sites deep in the brain.

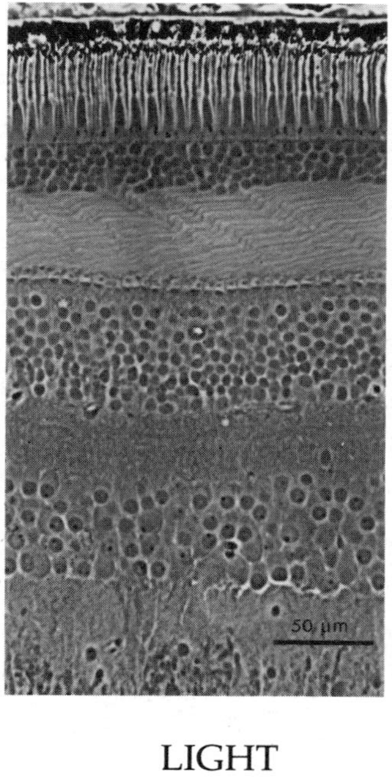

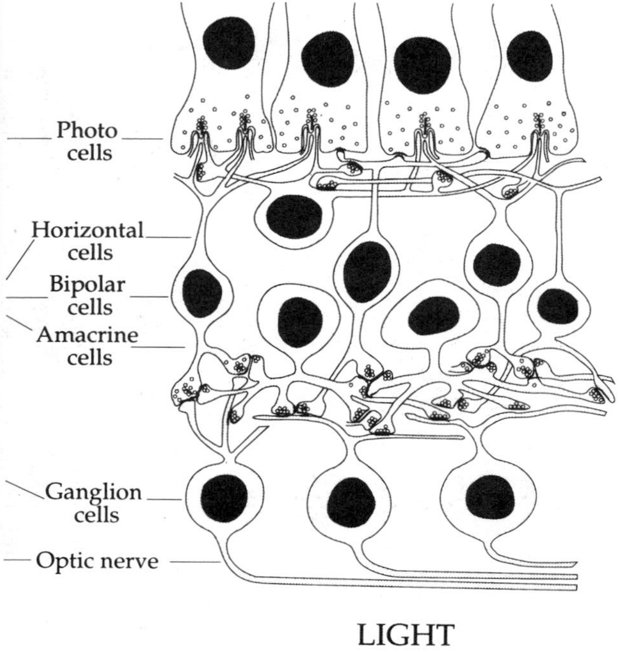

Photo cells

Horizontal cells

Bipolar cells

Amacrine cells

Ganglion cells

Optic nerve

LIGHT

LIGHT

50 μm

neurotransmitters produced by other neurons and delivered through a variety of synapse types.

When a neuron receives a jolt of neurotransmitter, its membrane voltage may be raised or lowered, depending on the synapse and neuron type. If the voltage is lowered enough by many signals, a kind of short circuit happens: the voltage suddenly collapses completely, and the collapse is propagated up the axon as a pulse. When the pulse reaches a synapse connecting the neuron to another, it triggers there a release of neurotransmitter from tiny sacks in its membrane. These diffuse across the synapse, eventually raising or lowering the voltage on the second cell. Meanwhile, the pumps in the first neuron work to restore the original voltage, and in a few thousandths of a second the cell is ready to fire again. The rate at which pulses are repeated encodes the intensity of the stimulus; anywhere from zero to several hundred pulses per second can be produced. Pulses are used for long-range communication, but closely spaced neurons, such as those found in the retina, often communicate simply by responding to each other's smoothly changing voltages. Besides the synaptic connections to other cells, many neurons and synapses have receptors for certain classes of free-floating neurotransmitters, delivered by the blood from other parts of the nervous system or other body organs, that inhibit or enhance the neuron's response.

At the nucleus of the neuron, slower genetic processes operate to manufacture neurotransmitter and to convey it down the axon to the storage sacks. The neuron's genetic machinery also packages energy, builds and repairs structure, and does all the other amazing things any cell must do to keep functioning. Fortunately for those of us working toward electronic imitations of the nervous system, most of this complexity is not directly involved in perceiving, acting, and thinking. Much of the neuron's mechanism is for growing and building an organism from inside out. Even its information-processing operations seem to be adapted from this evolutionary necessity, and it shows.

At this stage in computer technology, it is easier to keep the construction and repair machinery outside rather than inside the functional parts. Factories produce integrated circuits and assemble them into working hardware quite effectively. This removes a great deal of excess baggage from the final product. Moreover, because of their roundabout method of operating, neurons are quite slow;

they seem incapable of generating much more than 100 signals per second. These days, electronic switches, always vastly simpler and now smaller than neurons, can switch as fast as 100 *billion* times per second. The great speed advantage of electronics will allow us to get by with fewer electronic switches than the number of neurons in the human nervous system. Electronics is also exceptionally precise, allowing things to be done systematically and efficiently.

Now back to the human retina: what does it actually do? A rough and ready answer can be found if we compare the function of its five different cell types. At the outermost level is a network of neurons that respond to contrast, motion, other more specific features of the object under view. Connected to this neural network is a layer of light-detecting *photocells*. This type of cell is subdivided into cone cells, which together discriminate colors, and rod cells, which do not.

That light must pass through the neural network to get to the photocells is a peculiar feature of the vertebrate retina—one hit upon early in the evolutionary history of vertebrates and locked into place. The independently evolved retinas of the invertebrate octopus and squid have their photoreceptors up front. The awkward position of the vertebrate retinal nerve net has greatly limited its size, but strong selection pressure has enhanced its efficiency and function. Small differences in visual acuity or speed must often have had life or death consequences among our ancestors, and the retinal neurons are in a unique position to rapidly and comprehensively abstract the essentials from an image. The retina is thus likely to be an exceptionally efficient piece of vertebrate neural machinery.

After adapting to a particular overall light level, clusters of photo-cells create a voltage proportional to the amount of light striking them. This signal is received by two classes of cells, the *horizontal* cells and the *bipolar* cells. The horizontal cells, whose thousands of fibers cover large circular fields of photocells, produce a kind of average of their areas. If the voltages of all the horizontal cells were mapped onto a television screen, a blurry version of the retinal image would be displayed. The bipolar cells, on the other hand, are wired only to small areas and would provide a sharp picture on the TV. Some of the bipolar cells also receive inputs from nearby horizontal cells and then compute a difference between the small bipolar center areas and the large horizontal surround. Viewed on our TV, their picture would look much paler than the original, except at the edges of objects and patterns, where a distinct bright halo would be seen.

The bipolar cell axons connect to complicated multilayer synapses on the axonless *amacrine* cells. Each *ganglion* cell collects inputs from several of these amacrine synapses and produces a pulsed output, which travels up its long axon. Each amacrine cell connects to several bipolar and ganglion cells, and some of the junctions appear to both send and receive signals. Some amacrine cells enhance the "center surround" response; others detect changes in brightness in parts of the image. On the TV, some of these would show only objects moving left to right, while others would reveal other directions of motion. Each ganglion cell connects to several bipolar and amacrine cells and produces pulse streams whose rate is proportional to a computed feature of the image. Some report on high contrast in specific parts of the picture, others on various kinds of motion or combinations of contrast and motion.

The TV I have been referring to is not totally imaginary. Sitting next to me as I write is a TV monitor that often displays images just like those described. They come not from an animal's retina but from the eye of a robot. The picture from a TV camera on the robot is converted by electronics into an array of numbers in a computer memory. Programs in the computer combine these numbers to deduce things about the robot's surroundings. Though designed with little reference to neurobiology, many of the program steps strongly resemble the operations of the retinal cells—a case of convergent evolution. The parallel provides a way to measure the net computational power of neural tissue.

Cells and Cycles

The human retina has 100 million photocells, tens of millions of horizontal, bipolar, and amacrine cells, and a million ganglion cells, each contributing one signal-carrying fiber to the optic nerve. All this is packaged in a volume a half-millimeter thick and less than a centimeter square, 1/100,000 the volume of the whole brain. The photocells interact with their neighbors to enhance one another's output, and their great multiplicity appears to be a way to maximize sensitivity; even a single photon can sometimes produce a detectable response. The horizontal and bipolar cells and the amacrine cell synapses each seem to perform a unique computation. The bottom line, however, is that each of the one million ganglion-cell axons reports on a specific function computed over a particular patch of photocells.

To find the computer equivalent for such a function, we will first have to match the visual detail of the human eye in our computer equivalent. Simply counting photocells in the eye leads to an over-estimate, because they work in groups. External visual acuity tests are better, but they are complicated by the fact that the retina has a small, dense, high-resolution center area, the *fovea*, which can resolve details more than 10 times as fine as the rest of the eye. Though it covers less than 1% of the visual field, the fovea employs perhaps one quarter of the retinal circuitry and one quarter of the optic nerve fibers. Under optimal seeing conditions, as many as 500 distinct points can be resolved across the width of this central region. This feat could be matched by a TV camera with 500 separate picture elements, or *pixels*, in the horizontal direction. The vertical resolution of the fovea is similar, so our camera would need 500 × 500, or 250,000 pixels, in all—which, incidentally, just happens to be the resolution of a good-quality image on a standard television set.

But don't we see more finely than conventional TV? Not exactly. The 500 × 500 array corresponds only to our fovea, spanning a mere 5° of our field of view. A standard TV screen subtends about 5° when viewed from a distance of 10 meters. At that range, the scanning lines and other resolution defects of the TV image are invisible because the resolution of our eye is no better. At closer range, we can concentrate our fovea on small parts of the TV image to get greater detail, and this gives us the illusion that we see the whole screen this sharply. We don't; our unconsciously swiveling eyes simply zip the foveal area rapidly from one place on the screen to another. Somewhere, in an as yet mysterious part of our brain, a high-resolution image is synthesized, like a jigsaw puzzle, from these fragmentary glimpses.

So the foveal circuitry in the retina effectively takes a 500 × 500 image and processes it to produce 250,000 values, some being center-surround operations, some being motion detections. How fast does this happen? Experience with motion pictures provides a ready answer. When successive frames are presented at a rate slower than about 10 per second, the individual frames become distinguishable. At faster rates they blend together into apparently smooth motion.

Though the separate frames cannot be distinguished faster than 10 per second, if the light flickers at the frame rate, the flicker itself is detectable until it reaches a frequency of about 50 flashes per second. Presumably in the 10–50 cycle range the simplest brightness change detectors are triggered, but the more complicated neuron chains do not have time to react. Movie projectors avoid most of the flicker while keeping the frame rate reasonably low by using a rotating shutter to flash each frame more than once. Television does the same thing by scanning each frame twice, once with the odd numbered scan lines and once with the even. The peripheral parts of the retina have faster motion detectors than the fovea (presumably the better to notice fast-moving dangers coming from the sides), and many people can detect TV and movie flicker in the corners of their eyes.

In our lab at Carnegie Mellon we have often programmed computers to do center-surround operations on images from TV-toting robots, and once or twice we have written motion detectors. To get the speed up, we have spent much programming effort and mathematical trickery to do the job as efficiently as possible. Despite our best efforts, 10-frames-per-second processing rates have been out of reach because our computers are simply too slow. With an efficient program, a center-surround calculation applied to each pixel in a 500×500 image takes roughly 25 million computer calculations, which breaks down to about 100 calculations for each center-surround value produced. A motion-detecting operator can be applied at a similar cost. Translated to the retina, this means that each ganglion cell reports on the computer equivalent of 100 calculations every tenth of a second and thus represents 1,000 calculations per second. The whole million-fiber optic nerve, then, is a conduit for the results of 1 billion calculations per second.

If the retina's processing can be matched by 1 billion computer calculations per second, what can we say about the entire brain? The brain has about 1,000 times as many neurons as the retina, but its volume is 100,000 times as large. The retina's evolutionarily pressed neurons are smaller and more tightly packed than average. By multiplying the computational equivalent of the retina by a compromise value of 10,000 for the ratio of brain complexity to retina complexity, I rashly conclude that the whole brain's job might be done by a computer performing 10 trillion (10^{13}) calculations per second. This is about 1 million times faster than the medium-size machines that now drive my robots, and 1,000 times faster than today's best supercomputers.

Estimates like these are vulnerable to attack from many directions (see Appendix 1). After all, controversy flares when one merely compares similar electronic computers, whose internal operations are well understood and whose performance can be tested in detail. Hence it would be foolish to expect consensus opinion about a comparison of radically different systems executing dimly understood functions. Nevertheless, my estimates can be useful even if they are only remotely correct. Later we will see that a thousandfold error in the ratio of neurons to computations shifts the predicted arrival time of fully intelligent machines a mere 20 years.

Memory

Having settled on a 10-trillion-operation-per-second (10 teraops) computer as a sufficiently powerful host for a humanlike mind, we still have to decide how much memory to include. In 1953 the IBM 650 computer performed 1,000 instructions per second and was equipped with 1,000 "words" of memory, each able to store one number, or one instruction. In 1985 the Cray 2 ran at up to 1 billion instructions per second and was packed with up to 1 billion words of memory. This ratio, shared by most computers, of about one memory word for each instruction per second of speed was shaped by the market and probably indicates the size necessary to contain problems sufficiently large to keep a computer busy for seconds to hours at a time—rates comfortable for human programmers. If it had this ratio, a humanlike computer would require 10 trillion words of memory, about 10^{15} bits. (A *bit*, or binary digit, is a tiny unit of information that encodes a choice between two equal possibilities. Computer words today are between 16 and 64 bits long. Larger machines tend to have longer words.)

But is this number compatible with what is known about the nervous system? During the last decade Eric Kandel of Columbia University and others have studied the cellular changes that occur in the sea slug Aplysia when it is conditioned by irritating stimuli. They found that learning manifests itself as long-lasting chemical changes in individual synapses between neurons, changes that affect the strength of the connections to other neurons. Each synapse can store only one such strength, and then only with limited precision. If we assign 10 bits—enough to represent a number to three decimal places of accuracy—to each synapse, and if this storage method is substantially correct for larger nervous systems, then the 10^{15} bit "standard" memory of a humanlike computer should be able to contain the information encoded in the 10^{14} synapses of a human brain.

Comparative Computational Power and Memory
Some natural and artificial organisms rated by the measures of this chapter. Current laboratory computers are roughly equal in power to the nervous systems of insects. It is these machines that have hosted essentially all the research in robotics and artificial intelligence. The largest supercomputers of the late 1980s are a match for the 1-gram brain of a mouse, but at $10 million or more apiece they are reserved for serious work.

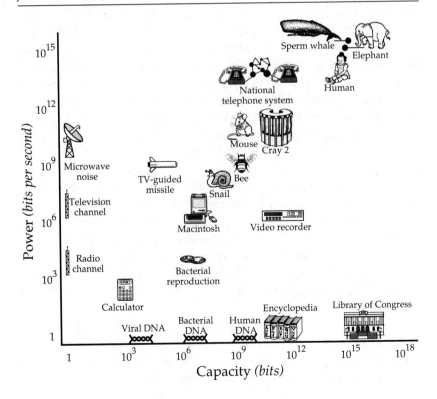

Comparing Computers

It is easy to see that computers are becoming more powerful, but by how much and how fast? Just when can we expect 10 teraops in a package sized and priced to fit in an autonomous robot? When I first approached this question, it seemed natural to rate electronic computers in operations per second, beginning with the first ones in the 1940s, and to project the resulting curve into the future. But there were complications. The machines came in many sizes, with prices ranging from tens of dollars to tens of millions of dollars. A given model could be equipped with many options, more memory, auxiliary processors, faster input and output, and so on. Recent machines are sometimes multiprocessors—multiple computers working in lockstep, or separately, sharing data. Machines had different instruction sets, so that an operation that took 10 instructions on one might be done on another in a single step. Some computers worked with numbers only 5 decimal places long; others handled 20 digits at a time. Also, the literature on the early electronic machines led me to their predecessors, computers built of telephone relays, the electromagnetic switches that had been perfected for telephone exchanges. Research on those, in turn, suggested yet earlier machines that calculated with motor-driven, and even hand-cranked, gears and cams. If these manual machines could somehow be compared with automatic computers, my curve could be extended back in time to the nineteenth century.

As a first step in devising a useful measure, I decided to cancel the size differences between machines by dividing each machine's processing power by its price, in constant dollars. This would give me an estimate of cost-effectiveness. For the mechanical calculators I added the human operator (valued at 100,000 1988 dollars—as if salary were a leasing cost) to the price, since to solve a problem a manual calculator needs a human to steadily enter numbers and operations and to write down results. This approach allowed the cost of purely manual calculation to also be measured—an unaided human clerk, whose effective capital cost is $100,000, can do about one calculation a minute!

Step two was to figure out just how factors such as speed, memory size, and instruction repertoire affected a machine's processing power. This was slippery. Computers today are often compared by measuring the running time of large sets of standard programs on each. This route was not open to me, since most of the machines I hoped to include in my curve no longer exist. I *did* know how long most of the machines took to add and to multiply two numbers, how many words of memory each had and the size of a word, and the approximate size of each machine's instruction repertoire. Processing power was to be the amount of computation done by the machine in a given time. If I could estimate how much computation each instruction accomplished, on the average, I would merely have to multiply by the number of instructions executed per unit time to get total power. So the problem reduced itself to estimating the work done by a single instruction.

Suppose a child's story begins with the words: *Here's my cat. It has fur. It has claws...* Pretty boring, right? Imagine, now, another story that starts out with: *Here's my cat. It wears a hat. It totes a gun...* Better. The second story seems more interesting and informative because its later statements are less likely—cats usually have fur and claws, but they rarely carry hats and guns. In 1948 Claude Shannon of MIT formalized such observations in a mathematical system that came to be known as *information theory*. One of its key ideas is that the information content of a message goes up as its likelihood, as measured by the recipient, decreases (mathematically, as the negative logarithm of the probability). A series of messages has maximum information content when it is maximally "surprising."

My measure of effective computation works the same way. Each instruction executed by a machine is like a message. The more predictable its sequence of instructions, the less useful work a machine is doing. For instance, a program that causes a computer simply to add one to a memory location once every millionth of a second is doing almost nothing of consequence. The contents of the memory location at any time in the future are known in advance. But even the best programs are limited in how much "surprise" they can introduce into a computation at every step. Each instruction can specify only a finite number of different possible operations and choose from a finite number of memory locations, each itself containing only a finite number of possibilities. These sources of surprise can be combined using the formulas of information theory to express the maximum information content of a single computer instruction.

I detail such a calculation in Appendix 2. The numbers vary from machine to machine and program to program, but I conclude there that a typical computer running an exceptionally efficient program

A Century of Computing

Cost of hardware for human equivalence *(1988$)*

$10^3 \quad 10^6 \quad 10^9 \quad 10^{12} \quad 10^{15} \quad 10^{18} \quad 10^{21}$

Human equivalence in a personal computer

Human equivalence in a supercomputer

GaAs integrated circuit

Microprocessor

Integrated circuit

Hybrid chip

Transistor

Vacuum tube

Electro-mechanical

Mechanical

Manual calculation

Cray 3
Mac II
Sun 4
Sun 3
VAX 8650
Vax 11/785
Cray 2
Vax 11/750
Macintosh
IBM PC
Sun 2
Sun 1
Cyber 205
Cray 1
VAX 11/780
PDP 11/70
Apple II
IBM 370/168
DEC KL 10
IBM 360/195
Nova
GE 635
SDS 920
DEC PDP 10
CDC 7600
IBM 360/75
IBM 360/65
CDC 6600
IBM 1130
B 5000
DEC PDP 6
DEC PDP 1
Atlas
IBM 7090
DEC PDP 1
LGP 30
IBM 1620
EDVAC
IBM 704
IBM CPC
Whirlwind
UNIVAC I
IBM 650
EDSAC
Zuse 5
Mark 2
SEAC
Mark 5
IBM SSEC
BTL Model 5
ENIAC
Colossus
Zuse 4
Zuse 3
BTL Model 3
Zuse 2
BTL Model 2
BTL Model 1
Burroughs 16
Zuse 1
Torres
National 3000
Monroe
IBM Tabulator
Babbage
Hollerith
Millionaire

$10^9 \quad 10^6 \quad 10^3 \quad 10^0 \quad 10^{-3} \quad 10^{-6} \quad 10^{-9}$

Computational power per unit cost *(bits/second/1988$)*

1900   1910   1920   1930   1940   1950   1960   1970   1980   1990   2000   2010   2020   2030

produces about 50 bits of surprise for each operation performed. If the computer can do 1 million operations per second, its maximum computational power is about 50 million bits per second. Expressed in these units, the computational power required in a human-equivalent robot is about 10^{14} bits per second.

Projections

The figure on page 64 plots the number of bits per second of computational power provided per (constant 1988) dollar of purchase price by a number of notable computing machines from 1900 to the present. Although numerous mechanical digital calculators were devised and built during the seventeenth and eighteenth centuries, only with the mechanical advances of the industrial revolution did they become reliable and inexpensive enough to rival manual calculation. By the late nineteenth century their edge was clear and the continuing progress dramatic. The vertical scale in the figure is logarithmic—equal steps represent tenfold increases in the ratio of performance to price. Since 1900 there has been a *trillionfold* increase in the amount of computation a dollar will buy. Each of the machines in the figure has a fascinating story; but since this book is not primarily a history of computation, I will restrict myself to a few highlights. *The Origins of Digital Computers: Selected Papers*, edited by Brian Randell, contains excellent first-hand accounts of many of these early machines.

Charles Babbage, of Cambridge, England, conceived the idea of an automatic program-controlled calculating machine in 1834, almost a century before anyone else. This "Analytical Engine" was to be a steam-engine-powered calculating behemoth of gears and shafts dealing in 50-digit decimal numbers. A rack of cogwheels was to store 1,000 such numbers, and a calculating unit was to be able to add two numbers in less than ten seconds and to multiply them in under a minute. The machine was to be controlled on the small scale by slowly rotating pin-studded drums such as those that still pluck the reeds in mechanical music boxes, and on a coarser scale by a stream of punched cards specifying memory locations and arithmetic operations to be performed with their contents.

In concept, the Analytical Engine contained all the elements of a modern digital computer. Babbage worked on it for the last 37 years of his life, but it was never completed. The enormous scale of the project and the tender state of the mechanical art (components were still typically hand fitted) made it unlikely that he could have succeeded in his lifetime. Precision interchangeable parts were much more common by the early twentieth century, and in 1910 Babbage's youngest son was able to demonstrate a working portion of the central calculating unit, although he did not complete the entire machine. I have included the Analytical Engine as a 1910 data point in my figure since it is likely that the machine could have been built at that time if a pressing need had arisen.

The other mechanical calculators in the chart were manually operated and were sold primarily to businesses for use by clerks and accountants, though some found uses in science. As mentioned previously, I included the "price" of the human operator in the cost of calculation for these. The early improvements in speed and reliability came with advances in mechanics: precision mass-produced gears and cams, for instance, improved springs and lubricants, as well as increasing design experience and competition among the calculator manufacturers. Powering calculators by electric motors provided a boost in both speed and automation in the 1920s, as did incorporating electromagnets and special switches in the innards in the 1930s.

Leonardo Torres y Quevedo, a Spanish inventor, demonstrated an electromechanical calculator in Madrid in 1919. Controlled by simple arithmetical commands entered at a typewriter keyboard, the "Torres Arithmometer" lacked a memory and was not fully automatic in the modern sense. Yet it was close, and it could have been converted to automatic operation by addition of a tape unit for entering commands, and made practical by addition of a unit to store and recall a handful of numbers.

Konrad Zuse independently invented the idea of programmed calculation as a young man in Germany in 1934; he built several large, automatic electromechanical computers in his parents' living room. The third machine in the series, built with backing from the German government and completed in 1941, was a complete, tape-controlled, binary floating-point (meaning it represented numbers in scientific notation, allowing for very small and very large numbers) computer with a 64-word memory. Zuse formed a company that sold improved models in the years following the war.

The Bell Telephone Labs (BTL) machines were built using telephone-exchange relay-switching techniques. The first two were modest internal projects, built to test antiaircraft gun directors. The third was a massive general-purpose, tape-controlled, automatic computer, intended as a commercial product. It was overtaken by much faster electronic machines and was never successful. The huge Harvard-IBM machines were of similar construction and suffered a similar fate. The era of general-purpose relay computers was over almost before it began.

One class of electromechanical machine had a longer history. The Constitution of the United States specifies that a national census must be taken every decade. As the country grew, these censuses took longer and longer to tally. The 1880 results were still being organized in 1887. It was obvious that without improved techniques the 1890 census would last beyond 1900. The Census Office held a competition for a better system. The winner was a young engineer named Herman Hollerith, who devised machinery that automatically counted holes in punched cards. Over the next half century Hollerith's invention evolved into a battery of "tabulating" machines that sorted and interleaved punched cards, duplicated them, printed on and from them, and did calculations with their contents. Hollerith's company grew into International Business Machines, which to this day represents 70% of the computer industry.

Electronic tube computers using radio and ultrafast radar techniques appeared as government-funded projects toward the end of World War II. The first commercially manufactured machine of this kind was the UNIVAC I, and its first customer was the Census Bureau, in 1951. By the end of the 1950s there were about 6,000 computers overall in industry, government, and universities. Their electronics was built around vacuum tubes, and they became known as the "first generation" of computers.

Beginning in about 1960, a second generation of machines began to appear that used newly developed transistors in place of vacuum tubes. They were smaller, more reliable, and cheaper and used less electricity than the vacuum tube computers, while providing more speed and memory.

By the late 1960s IBM began to introduce a third generation of machines using "hybrid integrated circuits." Dozens of tiny, unpackaged transistors and other electronic components were bonded onto wiring printed on ceramic chips the size of a thumbnail. Over the next several years these hybrid chips gave way to "monolithic" integrated circuits, in which dozens of components were etched directly into silicon chips a few millimeters square.

Integrated circuit technology developed rapidly, and by the mid-1970s a chip could contain thousands of components. A fourth generation of computers, whose heart was a handful of such chips, appeared but was quickly eclipsed by the microprocessor, a chip with tens of thousands of components that was, by itself, a complete computer. Progress was now so bewilderingly fast and multifaceted, with computers appearing in everyday devices such as microwave ovens, that the industry gave up on the generational nomenclature. (The last vestige was the Japanese Fifth Generation project, a research effort to develop artificially intelligent machines.) I am typing these words on a Macintosh II computer, a machine containing many chips with over a million components each, and a machine without a generation!

Human Equivalence in 40 Years

The progress documented in the figure on page 64 is remarkably steady despite radical changes in the nature of computing this century. The amount of computational power that a dollar can purchase has increased a thousandfold every two decades since the beginning of the century. In eighty years, there has been a *trillionfold* decline in the cost of calculation. If this rate of improvement were to continue into the next century, the 10 teraops required for a humanlike computer would be available in a $10 million supercomputer before 2010 and in a $1,000 personal computer by 2030.

But can this mad dash be sustained for another forty years? Easily! The curve in the figure is not leveling off, and the technological pipeline contains laboratory developments that are already close to my requirements. To a large extent, the slope of the figure is a self-fulfilling prophecy. Integrated circuit manufacturers have been aware of the trend since Gordon Moore, one of the inventors of the integrated circuit, noted in 1963 that the number of components on a chip was doubling each year. Computer makers have had similar observations, and new products in both of these related fields are designed with the trend in mind. Established manufacturers design and price products to stay on the curve, to maximize profit; new companies aim above the curve, to gain a competitive edge. The industry's success is one reason the success can continue—its enormous, and rapidly increasing, wealth supports more and better research and development of further advances. Also, the very computers that the industry makes are employed in the design of future circuits and computers. As they become better and cheaper, so does the design process, and vice versa. Electronics is riding these vicious cycles so quickly that it is likely to be the main occupation of the human race by the end of the century.

A key driver of both this decline in price and gain in performance is miniaturization. Small components simultaneously cost less and operate more quickly. Charles Babbage realized this in 1834. He wrote that the speed of his Analytical Engine, which called for hundreds of thousands of mechanical components, could be increased in proportion if "as the mechanical art achieved higher states of perfection" his palm-sized gears could be reduced to the scale of clockwork, or further to watchwork. (Try to imagine our world if electricity had not been discovered and the best minds had continued on Babbage's course. By now there might be desk- and pocket-sized mechanical computers containing millions of microscopic gears, computing at thousands of revolutions per second.)

To a remarkable extent the cost per pound of machinery has remained constant as the machinery has become more intricate. This is as true of consumer electronics as of computers (merging categories in the 1980s). The radios of the 1930s were as large and expensive as the televisions of the 1950s, the color televisions of the 1970s, and the home computers of the 1980s. The volume required to amplify or switch a single signal dropped from the size of a fist in 1940 to that of a thumb in 1950, to a pencil eraser in 1960, to a salt grain in 1970, to a small bacterium in 1980. In the same period, the basic switching speed rose a millionfold and the cost declined by the same huge amount. I cannot tell you exactly what developments will yield the additional factor of a million I project—such predictions are impossible for many reasons. Entirely new and unexpected possibilities are encountered in the course of basic research. Even among the known contenders, many techniques are in competition, and a promising line of development may be abandoned simply because some other approach has a slight edge. I *can* tell you that there are experimental components in laboratories today that improve on the best commercial components a thousandfold, at least in speed and size. Here is a short list of what looks promising today.

In recent years the widths of the connections within integrated circuits have shrunk to less than one micrometer, perilously close to the wavelength of the light used to print the circuitry. The manufacturers have switched from visible light to shorter wavelength ultraviolet, but this gives them only a short respite. X-rays, with much shorter wavelengths, would serve longer, but conventional x-ray sources are so weak and diffuse that they need uneconomically long exposure times. High-energy particle physicists have an answer. Speeding electrons curve in magnetic fields and spray photons like mud from a spinning wheel. Called *synchrotron radiation* for the class of particle accelerator where it became a nuisance, the effect can be harnessed to produce powerful beamed x-rays. The stronger the magnets, the smaller the synchrotron. With supercold superconducting magnets, an adequate machine can fit into a truck; otherwise it is the size of a small building. Either way, synchrotrons are now of hot interest and promise to shrink mass-produced circuitry into the submicron region. Electron and ion beams are also being used to write submicron circuits, but present systems affect only small regions at a time and must be scanned slowly across a chip. The scanned nature makes computer-controlled electron beams ideal, however, for manufacturing the "masks" that act like photographic negatives in circuit printing.

Smaller circuits have less electronic inertia; they switch faster and need lower voltages and less power. On the negative side, as the number of electrons in a signal drops, the circuit becomes more prone to thermal jostling. This effect can be countered by cooling, and indeed fast experimental circuits in many labs now run in supercold liquid nitrogen. One supercomputer is being designed to operate this way. Liquid nitrogen is produced in huge amounts in the manufacture of liquid oxygen from air, and it is cheap (unlike the much colder liquid helium). Uneven clumping of key impurities results in erratic component values as circuits get smaller, so more precise methods for implanting them are being developed. Quantum effects become more pronounced, creating new problems and new opportunities. Superlattices—multiple layers of atoms-thick regions of differently doped silicon made with molecular beams—are such an opportunity. They allow the electronic characteristics of the material to be tuned and permit entirely new switching methods, often giving tenfold improvements. Even more exciting are "quantum dot" devices that exploit the wavelike behavior of small numbers of electrons trapped in regions smaller than the electron wavelength.

The first transistors were made of germanium; they could not withstand high temperatures and tended to be unreliable. Improved understanding of semiconductor physics and ways of growing silicon crystals made possible faster and more reliable silicon transistors and integrated circuits. Newer materials are now coming into their own. The most immediate is gallium arsenide. Its crystal lattice impedes electrons less than silicon and makes circuits up to ten times faster. The Cray 3 supercomputer, scheduled to appear late in 1988, uses gallium arsenide integrated circuits packed into a one-cubic-foot volume, to top the Cray 2's speed tenfold. Other compounds like indium phosphide and silicon carbide wait in the wings. Pure carbon in diamond form is a definite possibility; it should be as much an improvement over gallium arsenide as that crystal is over silicon. Among its many superlatives, perfect diamond is the best solid conductor of heat, an important property in densely packed circuitry. The vision of an ultradense three-dimensional circuit in a gem-quality diamond is compelling. As yet no working circuits of diamond have been reported, but the odds improved in 1987 with reports from the Soviet Union, Japan, and, belatedly, the United States, of diamond layers up to a millimeter thick grown from microwave-heated methane.

The ultimate circuits may be superconducting quantum devices—not only extremely fast but highly efficient. Superconducting circuits have been in and out of fashion over the past twenty years. They have had a tough time because the liquid helium environment they required until recently is expensive, the heating/cooling cycles were stressful, and especially because rapidly improving semiconductors offered such tough competition. The newly discovered high-temperature ceramic superconductors could solve many of these old problems all at once. A superconducting transistor announced by Bell Laboratories early in 1988 is less than one twentieth of a micrometer in size and able to switch between on and off states in a picosecond (one trillionth of a second) on receiving an input signal of only one electron! A thousand microprocessors made of such switches would fit in the space occupied by one of today's microprocessor chips, and each would be a thousand times as fast. A thousand processors, each a thousand times faster than today's, would have just about the 10 teraops needed for human equivalence.

Farther off the beaten track are optical circuits that use lasers and nonlinear optical effects to switch light instead of electricity. Switching times of a few picoseconds, a hundred times faster than conventional circuits, have been demonstrated, but many practical problems remain. Finely tuned lasers have also been used with light-sensitive crystals and organic molecules in demonstration memories that can store up to a trillion bits per square centimeter.

Underlying these technical advances, and preceding them, are equally amazing advances in the methods of basic physics. One unexpected, and somewhat unlikely, device is the inexpensive scanning tunneling microscope that can reliably see, identify, and manipulate single atoms on surfaces by scanning them with a sharp needle. The tip is positioned by three piezoelectric crystals that stretch microscopically under the influence of small voltages. A gap of a few atoms in size is maintained by monitoring the current that quantum-mechanically tunnels across it. The tunneling microscope provides a secure toehold on the atomic scale, and big ideas about little atoms are being pursued by enthusiasts in both semiconductor and biotechnology laboratories.

Living organisms are clearly machines when viewed at the molecular scale—in them information encoded in RNA "tapes" directs protein assembly devices called ribosomes to pluck specific sequences of amino acids from the environment and attach them to the ends of growing chains of protein. Proteins, in turn, fold up in certain ways, depending on the sequence of their amino acids, to do many jobs. Some proteins have moving parts like hinges, springs, and latches triggered by templates. Others are primarily structural, like bricks or ropes or wires. The proteins of muscle tissue work like ratcheting pistons.

Today's biotechnology industry depends on modest manipulations of natural genetic machinery. The visionaries have more elaborate plans—nothing less than the fusion of biological, microelectronic, and micromechanical techniques into a single, immensely powerful, new technology. Computer-modeling techniques are slowly becoming powerful enough to allow new proteins to be designed and tested on display screens, much as conventional machine parts are often developed now. Such engineered proteins, as well as existing protein mechanisms copied from living cells, could be assembled into tiny artificial machines. Early products might be simple tailored medicines and small experimental computer circuits. Gradually, though, accumulated tools and experience would allow the construction of more elaborate machinery, eventually as complicated as tiny robot arms and equally tiny computers to control them. These would be small enough to grab individual molecules and hold them, thermally wriggling, in place. The protein robots could then be used as machine tools to build a second generation of even smaller, harder, and tougher devices by assembling atoms and molecules of all kinds. For instance, carbon atoms might be laid, bricklike, into ultrastrong fibers of perfect diamond. The entire scheme has been called *nanotechnology*, for the nanometer scale of its parts. By contrast, today's integrated circuit microtechnology has micrometer features, a thousand times bigger. Some things are easier at the nanometer scale. Atoms are perfectly uniform in size and shape, if somewhat fuzzy, and behave predictably, unlike the nicked, warped, and cracked parts in larger machinery. At the nanometer scale, the world is stocked with an abundance of components of absolute precision.

Atomic-scale machinery is a wonderful concept and would take us far beyond the humanlike point in computers, since it would allow many millions of processors to fit on a chip that today holds but one. Just how fast could each individual nanocomputer be? Quantum mechanics demands a minimum energy to localize an event to a given time: *Energy = h / time* where *h* is Planck's fundamental constant of quantum mechanics. Higher speeds require greater energy. Above the frequency of light, about a quadrillion (10^{15}) transitions per second, the energy reaches one electron volt, close to the energy of the chemical bonds holding solid matter together. Attempts to switch faster will tear apart the switches. A quadrillionth of a second, or femtosecond, is a million times faster than the nanosecond (billionth of a second) switching time of the fastest commercial computer components today, so a single nanocomputer might have a processing speed of a trillion operations per second. With millions of such processors crammed onto a thumbnail-size chip, my human-equivalence criterion would be bested more than a millionfold! That might seem to be enough, but I cannot help wondering if, just maybe, speeds beyond this "light barrier" are possible.

The physics world is a turbulent one, as theorists chase a goal that eluded Einstein: a single theory that encompasses all the types of particle and energy in nature. The front-runner these days is the superstring model; in it, particles are tiny loops of space knotted in six dimensions beyond the four of normal spacetime. Its variations predict a host of particles heavier than those making up atoms, some of them stable. Material made of such particles would be from a thousand times to astronomically denser and more tightly bonded than normal matter. Ultradense matter could, in principle, support switching operations much more rapid than the frequency of light. The benefits of miniaturization need not stop at the atomic scale! While ultradense matter that is stable on earth is just a speculation, vast quantities of similar stuff is known to exist in the tremendous gravity fields of collapsed white dwarf and neutron stars. Someday, our progeny may exploit these bodies to build machines with a million million million million million (that's 10^{30}) times the power of a human mind.

3 | *Symbiosis*

THE robot who will work alongside us in half a century will have some interesting properties. Its reasoning abilities should be astonishingly better than a human's—even today's puny systems are much better in some areas. But its perceptual and motor abilities will probably be comparable to ours. Most interestingly, this artificial person will be highly changeable, both as an individual and from one of its generations to the next.

But solitary, toiling robots, however competent, are only part of the story. Today, and for some decades into the future, the most effective computing machines work as tools in human hands. As the machinery grows in flexibility and initiative, this association between humans and machines will be more properly described as a partnership. In time, the relationship will become much more intimate, a symbiosis where the boundary between the "natural" and the "artificial" partner is no longer evident. This collaborative route is interesting for its powerful human consequences even if, as I believe, it will matter little in the long run whether or not humans are an intimate part of the evolving artificial intelligences.

We will begin our exploration of the symbiotic path with a history of its humble beginnings, the minimal interfaces of the early computers.

Stored Programs and Assemblers

ENIAC, the first general-purpose electronic digital computer, built in Philadelphia in 1946, was designed for hardwired control. A typical program involved thousands of wires connected by hand from point to point on large programming boards. "Writing" such a program was tedious; debugging one was daunting. Only a few were written before John von Neumann made a now-famous proposal. ENIAC had three large banks of dial switches, called *function tables*, intended for storing precomputed mathematical results needed during a calculation. They might be set up, for instance, with square roots or logarithms or with more specialized functions. Von Neumann suggested that these switches could be used in a different way, namely, to hold sequences of instructions, encoded as numbers, that would direct the machine's operation. The regular hardwired program would be set up, once and for all, in such a way that it could read these instructions from the function tables one after another and do what the numbers indicated. Thenceforth the machine could be programmed for new tasks simply by dialing commands into the function tables.

ENIAC
The rat's nest of wires at the left are the machine's original program-ming boards. The three banks of switches on the right were intended to hold mathematical function tables but were soon enlisted as a more convenient way to represent programs.

This new mode of programming was much easier and neater than the original scheme. Instead of a rat's nest of wires, a program consisted of neat columns of numbers. The numerical encoding system for machine operations came to be called *machine language*. It was one small step for user-friendly computers, a giant leap for computer organization.

All digital computers after ENIAC incorporated an expanded version of this *stored program* idea. Not only were programs represented as sequences of numbers, but the numbers were kept in the same memory used for calculations and could be loaded at high speed from input devices such as punched paper-tape readers. This unity of memory allowed a computer to modify its own program in midrun, an intriguing technique that was used extensively at first but is less common now. ENIAC used a large roomful of vacuum tubes to store fewer than fifty numbers, and could carry out about one thousand calculations per second.

ENIAC's successors were able to store entire programs in their working memory because new methods were invented for more densely and cheaply storing the electronic tally marks that constituted the memory. In some, a device much like a television picture tube was able to retain thousands of such "bits" as tiny areas of electric charge on its glass face. A sweeping electron beam could both sense and alter the contents of each area. In others, thousands of bits were encoded as a recirculating stream of acoustic pulses that traveled down a long column of mercury, to be sensed electronically at the end of their journey, amplified, and re-injected back into the head of the column. Another approach was to record the bits magnetically on the surface of a rapidly spinning drum or disk. Magnetic disks evolved into the bulk external storage devices still in use today, but they were too slow to survive long as the internal working memory of computers.

The most successful method turned out to be one that used tiny donuts of a specially developed magnetic material strung on the intersections of a net of fine wires. Each of these *magnetic cores* could store one bit, encoded as either a clockwise or counterclockwise magnetization. The direction of magnetization of a particular magnetic core could be changed by sending small currents through the horizontal and vertical wires that passed through it. Its previous contents could be determined with the help of another wire that zig-zagged through all the cores. Whenever any core's magnetization flipped from a clockwise to a counter-clockwise direction, or vice versa, a tiny pulse appeared on this wire. Magnetic core memories holding thousands and eventually millions of numbers were the main form of computer working storage for over twenty years, until they were overtaken in the mid 1970s by storage circuits much like those of ENIAC, but made of transistors instead of tubes and arrayed in thousands on tiny silicon chips.

Machine-language programming was a great labor saver. But increased memory, speed, and availability of computers soon lured their users to problems so large and complex that even machine language became unbearably tedious. A machine-language program consists of a sequence of instructions encoded as numbers. A few digits of each number, the *operation code*, specify the action the computer is to perform, say to add or to obtain the next instruction from elsewhere in memory. The remaining digits contain the *address* of one or more locations in memory indicating, in our examples, where to find the numbers to add or the next instruction. Converting problems originally formulated in terms of algebraic expressions like $x^2 + y$ to numerical codes that the computer could use was an unintuitive, slow, and error-prone process. Even worse, inserting a few additional instructions into a program might require shifting the location of much of the rest of the program and of number storage areas in memory, which in turn would require the alteration of the address parts of many, perhaps thousands, of instructions. The slightest error in the process could prevent the program from working. Programmers who became skilled at this exacting drudgery, punctuated by bursts of artful invention, were sometimes treated with the deference accorded to chess masters. But it did not long escape notice that computers themselves specialize in just this sort of exacting drudgery.

By the mid-1950s programmers were writing large programs whose function was to translate symbolic commands (like ADD X) into machine language while automatically assigning and keeping track of the location of variables and instructions. Machine-language purists complained that *assemblers*, as programs of this kind came to be called, lessened the precise control one should have over the workings of a computer and wasted computer time doing the translation. Even so, symbolic programming proved such a boon that writing machine language quickly became an extinct art. The path connecting humans and computers had widened again, allowing an increased flow of everyday traffic. The road enhancements also facilitated the passage of heavier road-building machinery.

Compilers and Operating Systems

Assemblers were a great help to professional programmers but were still very tedious for occasional computer users with particular problems to solve. Such end-consumers of computer power requested, and were granted, *high-level languages* that let compact mathematical notation, such as they used routinely in their work, substitute for longer, and more error-prone, assembler sequences. For instance, in a high-level language such as FORTRAN, which was one of the first and still survives today, $A \times X + B$ might stand in for the assembly sequence:

```
LOAD   A
MULT   X
ADD    B
```

Very complex programs called *compilers*, tours-de-force of programming when they were first written, translated lines of the high-level language into often long machine-language sequences. The user of a high-level language had little or no knowledge of the final program produced. Compilers used even more of the computer's valuable time than did assemblers and, lacking the cleverness and insight of human programmers, produced machine language that was larger and slower running than that produced when the same problem was written in assembly language. The disadvantages were great enough that, until recently, many critical applications still demanded assembly language.

But high-level languages had many advantages for the average user. They made programs easier to write, and many errors could be detected during the translation instead of causing obscure failures when the program was run. Since high-level languages were close in form to conventional mathematical notation, many computer non-specialists found them easier to learn. A transcendent feature was *transportability*. Unlike machine language or assembly programs, high-level programs did not reflect the detailed workings of any particular machine and so could be translated for entirely different computers. FORTRAN compilers, for instance, exist for essentially every computer ever built, and certain useful FORTRAN programs have survived the entire history of computers, leaping from one generation of machine to the next.

On the first digital computers, setting up a program, monitoring its progress, and cleaning up afterward was a strictly manual affair. The programmer could watch the action on banks of lights showing the internal state of the machine and could interrupt, examine, and alter it at any time, or run it step by step. This was quite convenient but expensive on machines whose time was valued at hundreds of dollars per hour. To minimize the time lost, computer companies in the late 1950s began supplying programs called *monitors, supervisors,* or *operating systems* to manage the flow of successive programs, read from punched-card readers or magnetic-tape units, through their machines. The early operating systems had a no-nonsense approach: one program at a time was run; if any problem was encountered it was automatically stopped, the contents of memory were printed out, and the next program was started. The *core dump* (as the memory printout was called, after magnetic core memory) was delivered to the programmer, who could diagnose the problem on his or her own time. Many programs spent only seconds in the computer, producing results that would occupy the programmer hours or days in preparation for the next run.

From the mid-1950s to the late 1960s this *batch* mode of computer operation was the rule, and a generation of computer users, especially users of IBM equipment, knew no other way. Some of the old timers, however, wistfully remembered the days of hands-on use. Not only was tracking down a programming bug easier with the computer as an ally, but it was possible to write programs that carried on a dialogue with the user. Interactive programs let the user and the computer act as partners, often with the user supplying insight and judgment and the computer providing prodigious calculation and memory. The problem was how to arrange this level of service without having the computer waste most of its time waiting for the human's next move.

Academic groups began work on a solution, an awesomely complex form of operating system that kept several programs, initiated from individual interactive terminals, going at the same time. Each user program would be allowed to run for a fraction of a second, then the *time-sharing* operating system would switch control to the next one and so on, eventually returning to the first. A human tied to any one of the active programs would not notice the fractional second interruptions, and it would appear to each user that he had a computer all to himself, albeit one that was a little slower than the raw machine. A program that became temporarily inactive waiting for user response would simply be skipped by the operating system, with little time wasted.

Like the assemblers and high-level languages before them, time-sharing systems caused considerable controversy in the computer community. With their need to keep the resources for several programs around at the same time, and to decide what to do next several times a second, they took a much larger cut of computer resources than batch operating systems. Time-sharing did offer efficiencies in return, however. With many diverse users active at one time, the various resources of a computer system—memory, disks, tape drives, printers, displays, and so on—could be kept busier than in single-program systems. More important, programmers could monitor the runs of their creations, quickly stopping them when, as often happens with new programs, they went awry. The most important efficiency gains, however, were for the customers rather than the machinery. Instead of waiting hours for a test run of a program on a batch system, a time-sharing user could watch the progress of a program, stop and modify it, and try again within minutes. This speed made possible a highly experimental, and somewhat Pavlovian, style of programming, characterized by quick punishment and reward cycles. A generation of proficient computer hackers was born.

The hackers, some spending much of their waking lives at computer terminals, quickly added to the basic capabilities of their favorite habitat. Entirely new uses for computers became the norm. Starting in the mid-1960s, users could communicate among themselves, live from terminal to terminal, or by electronic mail to be read and answered at leisure. They could engage in interactive text and video games, play sophisticated tricks on each other, share insights and programs, and in general experience a sense of community through the unwitting medium of their employers' machines. The community was enriched by the creation of many public artifacts in the form of computer files— community bulletin boards, witty quotes, technical hints, original writings, amusing programs, and interactive information-retrieval systems to help sort through all the richness. Some of the more sophisticated research systems also offered pictures and sounds from the machine.

The hackers' intensive and varied computer use called for efficient ways to find, start, and stop many different programs, to scan, read, and modify information files, to interact with other users, and to ask the computer to do many things automatically. In an evolutionary process, the command languages with which users at their terminals communicated with time-sharing operating systems were given these abilities. Designed by experts for experts, with layer upon layer of unplanned extensions, controlled by terse and powerful but often inconsistent, undocumented, and hard-to-remember incantations, these systems often exasperated the less experienced users. To a hacker the interface with the computer was instinctive, fast, and immensely powerful—almost anything was possible to one who could construct the right spells. But to those lacking the monomania required to keep up with the system's rapid and amorphous evolution, the interface was opaque, unhelpful, and very error-prone. The hackers got their comeuppance when dissimilar systems were hooked together by computer networks—even a hacker is a novice in another group's arcana.

A major success of the hacker era is the Unix time-sharing system. Created in the early 1970s by two young hackers at Bell Laboratories and extended by others at the University of California at Berkeley, it became, by the 1980s, the de-facto standard for larger computers. Unix is now working its way into upscale personal computers.

Menus and Icons

From time to time computer manufacturers tried to incorporate other hacker innovations into systems for their more staid customers. A major goal often was to make the language that invoked the many functions of the operating system as much like plain English as possible, in the expectation that this would greatly ease the burden on users already proficient in English. Such an effect had been observed in the scientific community when high-level languages that used more or less standard mathematical notation were introduced. Dissenters noted that, unlike mathematics, natural English is a poor language for precise descriptions; communities requiring precision invariably generate their own specialized conventions, mathematical notation being the most obvious example.

This time the nay-sayers were right. Experimental English-like interfaces (with long forgotten acronyms that tended to include the words "Plain" or "Simple" and "English") were researched during the late 1960s and early 1970s but were not very successful. It was (and is) not yet possible to embody true, general language understanding, with its underlying requirement of common sense and broad knowledge of the world, in a program. The actual systems failed to understand (or positively misunderstood) many offered phrases, so that using them was often a guessing game. It took skill to phrase requests so they would be properly interpreted by their complex but incomplete and largely undocumented language parsers. Learning a simple, consistent, special command code was easy by comparison.

Infrequent computer users (and specialists also) were better served by the invention of a much simpler device—the *multiple-choice menu*. In menu-driven systems, the major options are presented in a list, from which the user chooses one item. This choice may lead to a second menu outlining subsidiary characteristics, and so on, until the desired action is fully specified. Standard menu systems are not without shortcomings. It takes time both for the computer to print and for the user to read long menus, and the choices may not always be phrased in the most compatible manner. Menu systems therefore tend to be slower than specialist languages operated by experienced users. Also, it is difficult to capture really complex ideas in a game of twenty questions. Hybrid systems that once in a while prompt for essay-type answers sometimes manage this problem fairly well.

The speed problem in printing menus was easy, if a little expensive, to overcome: it required simply the installation of faster-displaying terminals. The new user's problem of quickly digesting long lists of unfamiliar options did not admit of so straightforward a solution. An excellent answer was explored and developed by hackers working in the exceptionally luxurious quarters of the Xerox Palo Alto Research Center (PARC). During the early 1970s this group developed expensive workstations, each with its own personal computer and large-screen display, capable of fast and fine graphic imagery. With good graphics and lots of personal computer power available, rather fancy interactions between user and computer became possible. For example, each computer had a hand-held device, called a *mouse*, which could be slid over a desktop like a hockey puck and which sensed this motion by means of a small rolling ball protruding from its base. The motion of the mouse was linked to the motion of a graphic arrow on the computer's display screen, and the user could thus point out any part of the screen to the computer. Pointing made menus more natural to use. Instead of typing the line label of a menu item, one could simply point to it to make the selection.

Small suggestive pictures attached to each menu line on the computer greatly eased the visual task of picking out the proper item and made the system partly language-independent. In later versions of this idea, the images, now referred to as *icons*, became the dominant representation, with words relegated to a small annotation. Eventually icons could be moved around the screen like objects and placed into other icons that functioned as containers or markers for physical destinations like printers. Icon interfaces proved effective and easy to use for novices and experts alike, probably because they tapped the nonverbal object manipulation skills of humans.

Providing each user with a separate computer had many implications, some of them confusing. Was it not a step backward from time-sharing? Several of the pioneers of time-sharing certainly thought so. The enthusiasts at PARC pointed out that, seen as a convenience for the user, time-sharing was seriously flawed. It immobilized the habitual user by tying him to a fixed terminal physically connected to a huge machine. Worse, the responsiveness of time-sharing systems had never lived up to their early promise; the number of users on the computer, and the overhead of the time-sharing system itself, invariably crept up until individuals found themselves waiting seconds, and even minutes, for the machine to respond to the most trivial requests. A small personal machine to handle the routine functions seemed plausible, given the rapidly declining cost of computers.

Alan Kay, the guru of the PARC group, originally envisaged a highly responsive, book-size personal computer (dubbed the Dynabook for its dynamic qualities) with a high-resolution color display and a radio link to a worldwide computer network. Much more than just a computer, the Dynabook would function as secretary, mailbox, reference library, amusement center, and telephone. This idea was, and still is, beyond the capabilities of the technology (though I am typing these very words into a book-sized portable computer as I sit in a hotel lobby).

The expensive PARC personal computers were desk- rather than book-size, had only black and white graphics, and offered less computing power than their designers desired. Nevertheless, they were a clear step toward the vision of the Dynabook. They were initially named Interim Dynabooks, but later, mercifully, acquired the more melodious name of Alto.

Xerox was slow to commercialize the discoveries of the PARC group, though in the late 1970s they did produce one expensive business workstation (named Star) that embodied many of the PARC ideas. After ten years of research, some of the PARC enthusiasts grew frustrated with the corporate sluggishness of Xerox and found a sympathetic ear in Steve Jobs, co-founder of the nearby Apple computer company. The result, a few years later, was the Apple Lisa and, later, its smarter younger sibling, the Macintosh, billed as the personal computer "for the rest of us." Though still far from a Dynabook in capability, it introduced a new way of computing to millions of people—and paved the way for a new phase of human–computer interaction. In the late 1980s almost all new operating systems for graphically oriented computers are being designed to present a Macintosh-style face to the user.

Magic Glasses

The graphical interface that makes the Macintosh and its imitators so much more pleasant to use than earlier machines demonstrates the value of engaging the sensory capabilities of humans in the dialogue between them and their machines. Alan Kay's Dynabook, though wonderful in many ways, would be unable to go much further than existing systems in this nonverbal direction because of its physical limitations, particularly the book-size viewscreen. As with a conventional book, portability is a key feature of the Dynabook; many of its proposed uses would vanish if it existed only in fixed locations like home or work. Can this portability be retained while at the same time the owner's sensory involvement is greatly expanded? In other words, can we imagine a computer that takes advantage of the mobility of humans, while allowing humans to take advantage of the superior memory, calculating power, and expanded communications range of the computer?

Of course we can—not in the form of a book, however, but in the form of a high-tech wardrobe. The key item of apparel is a pair of *magic glasses* (or, in the primitive stages, goggles or a helmet). Worn on the nose like conventional specs, these contain the following impressive array of instrumentation:

- High-resolution color displays, one for each eye, with optics that cover the entire field and make the image the computer presents appear to be focused at a comfortable distance. The glasses *may* have the ability to switch to a transparent mode.

- Three TV cameras. A high-resolution pair with forward-looking, wide-angle lenses is placed as close to the position of the eyes as possible, so that one can see where one is going when the cameras are connected to the corresponding display screens on the lenses. Perhaps a third very wide angle camera looks back to register most of the wearer's face, allowing the computer, and any videophone communicants, to monitor the wearer's facial expressions.

- Microphones and small earphones in the frames.

- A navigation system that accurately and continuously tracks the position and orientation of the glasses (and consequently the head of the wearer).

- A powerful computer that can generate realistic synthetic imagery, sound, and speech; understand spoken commands; and identify and track objects in the field of view of the cameras.

- A high-speed data link to a worldwide network of computers and electronic libraries, as well as to magic glasses worn by others.

It does not take an expert to recognize that this is a demanding set of requirements. Yet every one of these functions exists and is the subject of extensive, well-funded research. Squeezing it into a tiny package is also challenging but not implausible in this day of cassette-box Walkmans and pocket televisions. The glasses let the computer control what you see and hear, in response to your verbal requests, and to keep track of your wanderings; they also allow it to watch your hands moving, and your facial expressions, though they cannot control what you feel with your hands. Since touching is an important source of information about the world, it would be nice if a computer let us know what things feel like, even when they are not in our immediate environment. Enter *magic gloves*.

Like the glasses, the gloves are a tour-de-force of technological wizardry. Each finger of the glove contains a grid of elements that create patterns of pressure and temperature on the finger of the wearer. Physiological experiments have shown that realistic impressions can

Magic Glasses (Early Model)

Instrumentation in military aircraft—where instant access to naviga-tion, sensor, and weapons data is a life-or-death matter—is evolving into magic glasses. This model was developed in 1986 for a Boeing-Sikorsky experimental helicopter project. Data from the aircraft's instruments are projected into the pilot's field of view. Radar blips, for instance, are made to appear at the actual locations of the objects being tracked.

be created by this arrangement; for instance, uniform pressure and cold is interpreted as "My fingers are immersed in water." The gloves have motors that act on the joints of the fingers, so the wearer can feel resistance to motions. The same mechanisms permit the computer to monitor the position of the fingers.

Magic gloves, like magic glasses, have their limitations. They can generate resistance only to finger motions, yet manipulation involves movement of the whole arm. So imagine a *motorized coat* that could supply to the arm joints the sense of presence that the gloves give to the hand. Compared with the gloves, and especially the glasses, the coat is simplicity itself. But in early versions of the magic wardrobe, the coat may become a convenient repository for all the hardware that does not fit into the smaller and more complex glasses and gloves. It may even have to be attached to an immobile seat: a real "armchair."

Robot Proxy
This robot proxy was developed in 1986 at the Naval Ocean Systems Center in Hawaii. The motions of the operator on the left are copied by the robot on the right, and the images from the robot's camera eyes are delivered to the operator's (bulky) magic glasses. The operator has the subjective sensation of being in the robot's body.

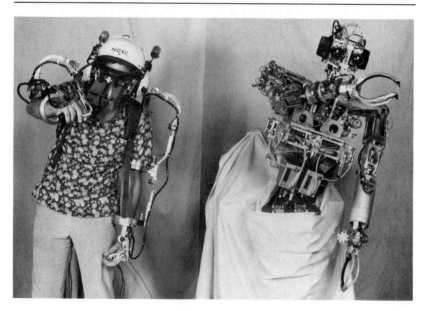

Later the coat will be as portable as a spacesuit, and eventually it will resemble a leisure suit.

As with existing computers, the wonderful hardware just outlined is only as useful as its software. The following sections present programs you might want to run when wearing your smart new outfit.

Finding Oneself

I don't know about you, but I often get lost. Lacking a good built-in sense of direction, I long for a pocket navigational aid that would not only tell me where I am (electronic maps now appearing in cars already do this) but would guide me to my destination, remember where I've been, and remind me of my grocery list when it notices I'm close to the store. The glasses contain a navigation system, probably a combination of a device that measures the distance to ground and orbiting beacons by radio and one that notes acceleration forces to deduce motion. However it is done, the navigator knows where you, the wearer of the glasses, are at any time.

Robin and the family have changed summer places since the last time you visited and now live somewhere in the backwoods. This is unfamiliar territory, and for once you really appreciate the glowing green line generated by the Yellow Brick Road program that guides you down the highway. The line veers into the right-hand lane, heading into a turnoff about a half-mile ahead. Just in case you didn't notice, a flashing right turn arrow hovers over the intersection, and the program whispers "Turn approaching, slow to 30" into your ear. A mile after the turn the asphalt becomes gravel, presaged by a color change in the guideline from green to a more cautious (and metaphorically correct) yellow. Later still, the line becomes red, and the road dirt. The program announces "Switching to private map" indicating that the Rand McNally database does not yet contain this little trail; the data comes from Robin's personal files. "Four miles to Robin's," the program offers, momentarily interrupting the old Beatles tune it's been playing.

It's getting dark. "Initiate phone—Robin," you say to the computer on your nose. This starts a telephone program that connects via the cellular/satellite network. The Yellow Brick Road program continues in parallel. "I'm almost there," you tell Robin. "Great. Dinner's just about ready. Do you like rice or potatoes?" "Both?" "Ok, see you!" "Bye." "Bye." The phone program terminates. The red guideline winds among the trees. The headlights are on but don't penetrate very far. Suddenly the car lurches to the left. You've hit something, perhaps a rock in the road, and now your front wheel is stuck in a muddy rut. Trying to drive out mires you more firmly. "Display map," you ask of the navigation program. The requested map hangs in the air before you. A little picture of your car on it shows you about two miles from your destination as the road meanders, though less than a half mile on a straight path through the woods.

"Hi Robin, me again. I'm in trouble." After considering the predicament, Robin suggests an adventurous dinner-saving course. "We can deal with the car in the morning. Last week Marty and I found a short-cut to near where you are. You should be able to walk here in under half an hour." Hardy soul that you are, you agree. The record of the path is transferred from Robin's computer to yours over the phone connection. Your navigation program links the short-cut route with the road data, and the original guideline is replaced by one that runs down the road for a distance, then snakes off into the woods. You lock your car and follow the line.

It is very dark, so you activate the night-vision program—the cameras in your glasses are run at maximum sensitivity and deliver enhanced images to the screens, added to the output of the navigator. You see that the red guideline follows a faint trail through the woods. Except for a few scratches, the walk is uneventful. The meal is warm and delicious.

Going Places

Transportation and communication have improved awesomely in the last 500 years. Yet geography, while no longer being the major determinant of commerce, still restricts how and with whom most of us conduct our affairs. The differences between transportation and communication will become less distinct as we become more able to project our full awareness and skills to remote locations. The ease of such projection will allow common interest to be the primary spur to association. The magic wardrobe can be used to visit both real places in the world and "unreal" locations deep inside computer simulations.

The most obvious form of remote presence involves a physical robot proxy—a distant robot that you control via the global communications network. The magic glasses allow you to see through the robot's eyes, the coat and gloves permit you to feel, gesture, and act through the robot's manipulators, and foot controls on your armchair let you drive the robot around. By renting proxies at remote locations, you can visit, talk, and work at widely scattered projects without leaving the comforts of home and without incurring the physical risks of dangerous locations or the boredom of long trips.

A twist on the proxy idea is a *human* proxy. Someone traveling to an interesting locale and wearing a special kind of magic wardrobe would be able to transmit their view, sound, touch, and perhaps odor impressions to the armchair traveler. In its simplest form, the link is one way, from proxy to passive observer. Such one-way communication can be recorded and played back at will—giving us a plausible form of the "feelie" extension to the "movie." Alternatively, the connection could be two-way, with motor actions and sense impressions, suitably edited by a clever program, transmitted bidirectionally. Inexperienced participants are likely to find themselves as amusingly uncoordinated as in a three-legged race, but training may permit an intimate kind of teamwork, with remote expertise being brought to bear at the location of the problem. Most of the time the "field agent" would be in control of the movements, while the armchair participant would watch, listen, feel, and give advice. But when the task called for a manual skill better known to the stay-at-home, the field agent would relax and allow the remotely controlled suit motors to do the job—as if possessed by a spirit.

A proxy meeting need not be in the real world—many things can be done better in computer simulation. Computer generated "unreal estate" has no intrinsic limits either in extent or in physical properties. It is a world where magic is routine. Today's computer screens allow peeks into this world—car designers examine future models, physicists view the interior of nuclear explosions, and Macintosh users rearrange their files on an unreal space called a "desktop." With a magic wardrobe, we will be able to go boldy into such worlds and explore them from the inside.

Your (modest) dream home is finally becoming a reality. The preliminaries with the architect were done weeks ago, and an eagerly anticipated call arrives: "Hi there. I've got a tentative design. Do you have time to look it over?" "You bet!" "Ok, lets switch to the site. Initiate Scene M5." After a few seconds the computer in your glasses prompts "Allow scene change?" "Allow," you confirm.

A pretty good rendition of the lot and its neighborhood surrounds you. "I thought we'd put the house over here. That gives a nice front and rear yard, and room for driveway and garage on the right." An outline of the house appears on the ground. "Hallway, guest washroom, living and dining room, kitchen, and stairwell on the first floor." A labeled floor plan shows up in the outline. You presume the architect's view of the scene is more complicated than your own and includes display controls. "Lets put up the first floor." The floor plan sprouts walls. Your point of view becomes higher, and you see a second floor plan on top of the truncated building. "Two bedrooms, two baths and an office at this level." The second floor grows into place. "The third floor is attic, with potential for two bedrooms and a bath." The roof

Unreal Estate—The Road to Point Reyes

This scene was synthesized at the Lucasfilm computer graphics division in 1985 from an underlying three-dimensional computer model. Magic glasses and enough computing power will allow us to stroll through fantasy worlds like this. © 1986 Pixar.

finishes the assembly, and a bit of landscaping is added. Drifting back to ground level and around the house, you take in the scene. "Could we try that in brick?" The walls change from cut stone facing to brick. "The back yard looks a little small. Could we move things forward?" "We can't get too far out of line with the neighbors, but I think I can give you 15 feet." The house slides forward. "That's better. Let's go inside."

The front door swings open. You note a light switch on the right and reach to flip it. A stick-figure caricature of a hand connects, and the hall is illuminated. The view from the living-room windows is not inspiring. "Could we have a bay window here, maybe, instead of this one?" "Hold on a sec, I'll have to set that one up." You drift into the kitchen, which looks pretty spacious, then examine the dining room. "Window's ready." The living room looks better with sunlight streaming in the bay window. "What season do you have set?" "Realtime. Let's cycle through a year." The lighting changes through morning and afternoons in all four seasons and the winter scene is particularly cheery. After a quick tour of the upstairs, with a few wall color changes, you bid goodbye. "I'll leave a copy. You'll be able to make cosmetic changes and fiddle with the furniture; don't worry about ruining the design—the program will prevent you from doing anything silly." "The family will be thrilled this evening. Talk to you later."

Bare-Hands Programming

Skilled practitioners in many fields report that they *see* or *feel* the object of their work as they think about it. This is not a great surprise in occupations that concern physical objects or situations— sculpture or sports, say, or mechanical design. It is less expected in supposedly abstract fields like music, language, mathematics, or theoretical physics. Yet Einstein, for instance, reported that he could often feel the meaning of his equations in his arms and his body as if they were solid objects.

As I suggested in Chapter 1, the large, highly evolved sensory and motor portions of the brain seem to be the hidden powerhouse behind human thought. By virtue of the great efficiency of these billion-year-old structures, they may embody one million times the effective computational power of the conscious part of our minds. While novice performance can be achieved using conscious thought alone, master-level expertise draws on the enormous hidden resources of these old and specialized areas. Sometimes some of that power can be harnessed by finding and developing a useful mapping between the problem and a sensory intuition.

Although some individuals, through lucky combinations of inheritance and opportunity, have developed expert intuitions in certain fields, most of us are amateurs at most things. What we need to improve our performance is explicit external metaphors that can tap our instinctive skills in a direct and repeatable way. Graphs, rules of thumb, physical models illustrating relationships, and other devices are widely and effectively used to enhance comprehension and retention. More recently, interactive pictorial computer interfaces such as those used in the Macintosh have greatly accelerated learning in novices and eased machine use for the experienced. The full sensory involvement possible with magic glasses may enable us to go much further in this direction. Finding the best metaphors will be the work of a generation; for now, we can amuse ourselves by guessing.

The familiar landscape of the top level of your file system lies ahead. In the foreground, on a grassy green meadow, are variously sized, colored, and shaped boulders labeled "Budget," "Drawings," "Games," and so on. In the fog-shrouded distance are large hills emblazoned "Oxford English Dictionary" and "Encyclopedia Britannica." Two knocks on the "Space" boulder cause it to expand and to open a portal in its side. What might be taken for an asteroid belt is visible through the portal. One of the rocks floating in the blackness is labeled "Skyhooks." You drift up to it, knock twice, and enter. A pretty blue and white earth, and some less pretty bits of variously shaped debris, greet you. This is an unfinished project, and some of your less successful experiments have yet to be laid to rest.

Today's problem is to develop a simulation of a long and strong cable orbiting the earth. The cable has mass and a certain stretch. It can be approximated (you've learned) by stringing together large numbers of simple springs and even simpler weights. A simple spring joins two points and exerts a force on them proportional to the amount of elongation from a rest length. A mass has a position and a velocity that changes in proportion to an applied force in accordance with Newton's three laws of motion. The formula for a spring is $F = K (I - I_o)$ where I is the spring's current length and I_o is its unstretched length. K is the spring constant: a larger K makes the spring harder to stretch. F is the force exerted by the spring on its endpoints. This relation is among the debris that litters the landscape. You begin by choosing some components you've constructed in the past. A point is imaged as a small black dot that hangs in space (internally it has three numbers giving its X, Y, and Z coordinates, but that was yesterday's concern). Tapping on the point and saying "Duplicate!" gives you a second one. These will represent the two ends of the spring.

You fetch a length arrow; it looks like a line with an arrowhead at each end and a number (its length) in the middle. Fasten its two heads to your two points and it calculates the distance between them. Taking it for a spin, you grab one of the points and move it around. The arrow follows the point and the length number changes obediently. Tapping it you say "Call this I." The dimension changes to the symbol I. "Attach spring formula," you command, and a copy of the formula springs from the landscape, settles nearby, and begins to respond to the distance between the points. Slots for the variables K and I_o appear, and you give them values. "Vectorize" is another prepared component; given a pair of points and a simple magnitude, it gives direction to the quantity, that is, the direction of the line joining the points. This is attached to both points, in opposite directions. You command "Vectorize F." The points now exert the spring force, though they remain fixed. When you reach to grab one or the other, it tugs on your hand—the farther you go, the harder it pulls back toward the fixed position of the other point. Only a few kinds of quantity can be directly experienced this way. Position, color, and temperature are others. In many applications it is helpful to translate more abstract measurements into palpable ones.

You add mass to the endpoints. This allows them to move independently, under control of momentum and applied forces, such as the spring force. With its ends released, the spring vibrates. The vibration does not diminish until you add a damping term to the force equation that diminishes the force, depending on the rate of change of I. Now the spring behaves reasonably, and you sproing it a few times for fun. Invoking the compiler converts the spring into a single object and greatly improves the efficiency of the underlying program. You edit the spring's image to make it look like a stretchy coil, with black disks representing mass on the ends. A dozen duplicates of the spring strung together end-to-end make a rather stretchy rope. Your simulation is off to a good start, but it's lunchtime. After lunch you'll make a longer section, alter the parameters in the various parts, and instrument it, perhaps by plotting the stretch of the various sections in a graph. Then you'll turn on the earth model's gravity and put the string into orbit around it and watch what happens.

Elementary Physics

Socrates, whose lessons were recorded for us by his student Plato, wrote no books himself. He seemed to think writing was a bad idea, since it allows its users to put on a show of knowledge by looking things up, without really knowing anything; the very capacity to remember, and to think about the memories, was jeopardized. Furthermore, an argument presented in a book provides no outlet for disagreement, unlike a person, with whom one can argue or obtain clarifications. Both objections have merit. Book knowledge is certainly dry and static compared with active knowledge in a clever person's mind. The invention of printing greatly aggravated the effect. Yet books have a reach, capacity, and permanence much greater than any person's memory, and these properties have made modern civilization possible.

Before the age of printing, books were expensive, laboriously hand-made items found in a few widely scattered libraries. Private copies were hard to obtain, and scholars found it necessary to memorize whole volumes. Artificial aids to memorization were valued and sometimes jealously guarded from theological, political, and commercial rivals. A very effective memory technique, developed into countless variations through the Middle Ages, was The Walk. A large location, perhaps a cathedral with many rooms, was remembered or imagined. A book or lecture to be memorized would be recited while at the same time the structure was mentally traversed. Each room of the cathedral, or portion of a room, would serve in the mind's eye as a repository for a section of the text, perhaps marked by some object that reminded one of the topic. The task was thus broken into manageable chunks; each location required only a moderate amount of remembering. The entire piece could be reconstructed by once again mentally walking through the building, visiting the rooms one by one, with the mental images so generated bringing to mind the associated portions of text.

The Walk may be so effective because it maps the new cultural need to memorize large quantities of speech into the much older survival skill of remembering where we saw or left various things. Recalling the location of a food source, shelter, a danger, a friend or foe, or simply a landmark once seen in the course of a journey has clear everyday benefits and is something many of us do naturally and quite well. At least a portion of our memory is likely organized in an approximately geographical way to facilitate this kind of recall.

A lesson delivered through an advanced edition of the magic wardrobe can simultaneously be as responsive as a personal dialogue, as permanent and available as a printed book, and in resonance with natural skills in a way that exceeds any existing method.

The "Gravity" portal opens onto a brightly sunlit pastoral scene. A tree-lined country road winds into distant hills, fluffy clouds dot the sky, birds are chirping somewhere. A few of the trees bear apples, and from time to time one falls to the ground. Some distance down the road a bewigged figure comes into view, sitting under one of the apple trees. You recognize Sir Isaac Newton. He looks just like he did in the "Laws of Motion" chapter.

"Greetings, young friend," says Sir Isaac. "I've been puzzling over the nature of the attraction of the earth for various objects. This apple, for instance, tugs at the hand with a certain force." He hands you the apple; sure enough, it has weight. "An apple with twice the substance pulls twice as strongly." The apple gets bigger and heavier. "The great Galileo observed that, when released, an object falls toward the ground with a constantly increasing velocity, independent of its weight." Galileo's demonstrations with falling balls in "Laws of Motion" come to mind. "Yes, yes, get on with it." Newton, with a slight frown, continues. "We can conclude that each particle of an object is attracted to the center of the earth with a force proportional to its mass. Does this attraction change with distance from the earth? One can conjecture that the influence extends to great distances and holds the Moon in its monthly circuit. If the same laws apply to celestial bodies as to the mundane, then our studies on the motion of objects indicates that a force in the direction of the earth's center suffices to bend the moon's path. Yet the required force is almost 4,000-fold weaker, per particle of mass, than is experienced by the apple you hold."

As he speaks the ground swells at a fantastic rate, and you, Newton, and the tree are on the summit of a hill rising like a rocket. "Consider the path of an object thrown horizontally from a great height; your apple, perchance." Taking the hint, you launch the apple with a smart upperhand throw. (In the real word, the motors in your jacket and gloves hum momentarily as they resist your moving arm, simulating the forces of the apple's inertia). The apple arcs slowly toward the ground and strikes near the horizon. The hill has stopped growing, but you are very high, and the spherical shape of the planet is evident. You can make out several continental outlines. This is obviously a miniature scale model of the earth. Sir Isaac hands you another apple and recommends a harder throw. It arcs beyond the horizon, in a curve almost paralleling the ground. You hear a splat through the ground under your feet. A yet harder launch results in no impact at all, and after a while the apple whizzes past your head from behind and goes round once again. A miniature side view of you, hill, earth, and apples makes all this clearer; each launch traces out an ellipse that returns to its starting point unless it intersects the ground first. Newton recalls Kepler's laws of planetary motion and claims they hold for the apples only if the attraction drops as the square of the distance from the planet's center. You're skeptical, so the two of you experiment with other rules. Some cause the apples to trace out nonrepetitive patterns. Those that do give ellipses violate Kepler's second law, that the line joining the planet center to the orbiting body sweeps out equal areas in equal times. After a while your throwing arm gets tired, and you say you're convinced.

But sometimes your skepticism leads to questions that stump your host. You remember Newton once responding, "A curious puzzle. Let me ponder it awhile." Several visits later he came puffing after you with the answer, coattails flying, one hand holding down his wig and trailing a cloud of dust. (You presume the book's software, unprepared to answer the question the first time, had issued a message about it to the book's authors. The authors then created entries in the book's database that allowed your pending query, and any similar ones Isaac encounters in future, to be answered.)

The hill shrinks back to flatness, and you're on the road again. Next stop is a pasture where some of the more formal parts of the lesson will be discussed. A half dozen exotic creatures are already gathered there. Many people the world over are reading this book, and the world network makes it possible for those who wish to associate to be mutually aware of each other during the course of the study. In such associations most people take advantage of the freedom of the simulation to assume forms different from their physical bodies, for anonymity and whim. Your group has a Wolf, Floating Eye, Tin Man, Giant Butterfly, Dragon, and Small Tank. You yourself appear to them as a rather stylish Dwarf with axe and tasseled hat. Some former classmates who began this physics book with you are no longer in your cohort because they sped ahead or fell behind your pace or took a different turn at a subarea branchpoint. From time to time you pick up new traveling companions as subcategories re-merge. The whole world is divided into overlapping "villages of common interest" of this kind. The group sizes range from two to several thousand. Often, of course, it's good to walk the paths of learning and entertainment in solitude. Among other advantages, the action can be better individually tailored, since there are fewer constraints.

After the lesson you glance farther down the road. In the distance is a railway platform with a stopped passenger train. Looking carefully, you note in the window of one of the cars the somewhat disheveled profile of the world's most famous scientist. But you're tired, and so you disconnect for the day. Relativity can wait for tomorrow.

4 | *Grandfather Clause*

WHAT happens when ever-cheaper machines can replace humans in any situation? Indeed, what will I do when a computer can write this book, or do my research, better than I? These questions have already become crucial ones for many people in all kinds of occupations, and in a few decades they will matter to everybody. By design, machines are our obedient and able slaves. But intelligent machines, however benevolent, threaten our existence because they are alternative inhabitants of our ecological niche. Machines merely as clever as human beings will have enormous advantages in competitive situations. Their production and upkeep cost less, so more of them can be put to work with the resources at hand. They can be optimized for their jobs and programmed to work tirelessly.

As if these technological developments were not threatening enough, the very pace of innovation presents an even more serious challenge to our security. We evolved at a leisurely rate, with millions of years between significant changes. Machines are making similar strides in mere decades. When multitudes of economical machines are put to work as programmers and engineers, presented with the task of optimizing the software and hardware that makes them what they are, the pace will quicken. Successive generations of machines produced this way will become smarter and less costly. There is no reason to believe that human equivalence represents any sort of upper bound. When pocket calculators can out-think humans, what will a big computer be like? We will simply be outclassed.

So why rush headlong into an era of intelligent machines? The answer, I believe, is that we have very little choice, if our culture is to remain viable. Societies and economies are surely as subject to competitive evolutionary pressures as are biological organisms. Sooner or later the ones that can sustain the most rapid expansion and diversification will dominate. Cultures compete with one another for the resources of the accessible universe. If automation is more efficient than manual labor, organizations and societies that embrace it will be wealthier and better able to survive in difficult times and to expand in favorable ones. If the United States were to unilaterally halt technological development (an occasionally fashionable idea), it would soon succumb either to the military might of unfriendly nations or to the economic success of its trading partners. Either way, the social ideals that led to the decision would become unimportant on a world scale.

If, by some unlikely pact, the whole human race decided to eschew progress, the long-term result would be almost certain extinction. The universe is one random event after another. Sooner or later an unstoppable virus deadly to humans will evolve, or a major asteroid will collide with the earth, or the sun will expand, or we will be invaded from the stars, or a black hole will swallow the galaxy. The bigger, more diverse, and competent a culture is, the better it can detect and deal with external dangers. The larger events happen less frequently. By growing rapidly enough, a culture has a finite chance of surviving forever. In Chapter 6 I will fantasize about schemes that would allow an entity to restructure itself so as to function indefinitely even as its universe ended.

The human race will expand into the solar system before long, and human-occupied space colonies will be part of that expansion. But only by a massive deployment of machinery can we survive on the surfaces of other planets or in outer space. The Apollo project, for example, put people on the moon for a few weeks for $40 billion, whereas the Viking landers functioned on Mars for years, at a cost of only $1 billion. If machines as capable as humans had been available for the Viking project, they would have been able to gather far more information about Mars than people were able to gather about the moon, simply because machines can be constructed to function comfortably and economically in unearthly environments.

Outer space is already a profitable arena for the owners of communications satellites. As transportation costs decline, other activities will start to pay. Space factories using raw materials purchased from earth or from human space outposts will be processed by human-supervised machines and sold at a profit. The high cost of maintaining humans in space ensures that there always will be more machinery per person in a space colony than there is on earth. As machines become more capable, the economics will favor an ever higher machine-to-people ratio. Humans will not necessarily become fewer at this stage; the machines will just multiply faster, becoming ever more competent with each new generation. Imagine the immensely lucrative robot factories that could be built in the asteroids. Solar-powered machines would prospect and deliver raw materials to huge, unenclosed, automatic processing plants. Metals, semiconductors, and plastics produced there would be converted by robots into components that would be assembled into other robots and into structural parts for more plants. Machines would be recycled as they broke. If their reproduction rate is higher than the wear-out rate, the factories will grow exponentially, like a colony of bacteria, on a Brobdingnagian scale. A harvest of a small fraction of the output of materials, components, and whole robots could make investors incredibly rich.

Eventually humans, whether workers, design engineers, managers, or investors, will become unnecessary in space enterprises, as the scientific and technical discoveries of self-reproducing superintelligent mechanisms are applied to making themselves smarter still. These new creations, looking quite unlike the machines we know, will explode into the universe, leaving us behind in a cloud of dust.

Robot Bushes

The human world has been shaped by human hands, which are still our most effective general-purpose tool. Yet many useful and easily described tasks are beyond human dexterity (pull tightly on both ends of the string, while holding the knot between your fingers, lift the bundle, wrap the ends around it tightly four times...). If such actions are attempted at all, it is with varying degrees of success using special tools and fixtures.

It is unlikely that our superintelligent descendants will be satisfied with mere stumpy fingers. Consider the following observations. Worms and other animals shaped like balls or sticks are unable to manipulate or even locomote very well. Animals with legs (a stick with smaller, movable sticks) locomote quite well but are still clumsy at manipulation. Animals like us, with fingers on their legs (sticks on sticks on a stick), can manipulate much better. Now generalize the concept—a robot that looks like a tree, with a big stem repeatedly branching into thinner, shorter, and more numerous twigs, ultimately ending in an astronomical number of microscopic cilia. Each intermediate branch would be able to swing forward and backward and side to side while its top, where the next smaller branches are attached, rotates on the branch axis. Possibly the branch could also change its length like a telescope—the number of motions

A Robot Bush

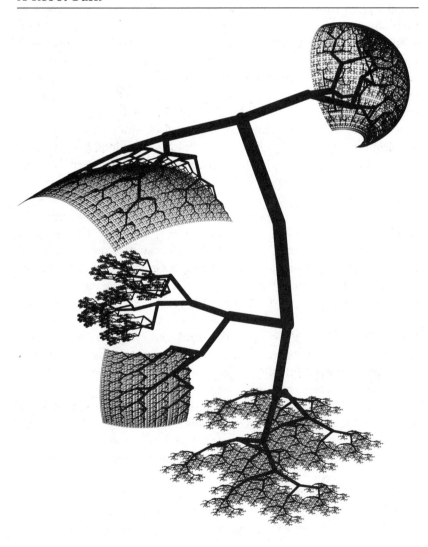

of each branch can be traded off for more levels. Each joint would have sensors to measure its position and also the force it exerts. Although made of branches, each with a rigid mechanical character, the overall structure would have an "organic" flexibility because of the great multitude of ways its parts could move.

A robot of this design could be self-constructing. Tiny bushes, only a few millionths of the weight of the final device, would be "seeded" to start the process. These would work in groups to build the next larger sprigs from available raw materials, then join themselves to their constructions. The resulting larger bushes would join to build even larger branches, and so on until a small crew (of large members) met to assemble the stem. At the other end of the scale, a sufficiently large bush should be able to organize the necessary resources to build the tiny seeds to start the process all over again (or simply to repair or extend itself). It could make the smallest parts with methods similar to the micromachining techniques of current integrated circuitry. If its smallest branchlets were a few atoms in scale (with lengths measured in nanometers), a robot bush could grab individual atoms of raw material and assemble them one by one into new parts, in a variation of the nanotechnology methods mentioned in Chapter 2.

To make things more concrete, we can do an actual design. Suppose that the basic structure is a large branch that splits into four smaller ones, each half the scale. If we start with a stem a meter long and ten centimeters in diameter and carry the branching to twenty levels, the bush will end in a trillion tiny "leaves," each a millionth of a meter (a micron) long and a tenth of that in diameter. Because of their much smaller weight and size, the leaves can move a million times as fast as the trunk. Let's say that the trunk can wiggle back and forth once per second; so the leaves will vibrate a million times per second. If the bush folds itself into a tight bundle, its cross section will be approximately constant, and it will be two meters long. The trunk will then occupy half that length, the second level half the remaining length, and so on. Unfolded, umbrellalike, it would spread into a disk a little under two meters in diameter, thick but sparse near the center, and thinner at the edge, with smaller gaps that taper off to micron spaces.

If each joint can measure the forces and motions applied to it, we have a remarkable sensor. There are a trillion leaf fingers, each able to sense a movement of perhaps a tenth of a micron and a force of a few micrograms, at speeds up to a million changes per second. This is vastly greater than the sensing ability of the human eye, which has a million distinguishable points that can register changes at most a hundred times per second. If our bush puts its fingers on a photograph, it will "see" the image in immense detail simply by feeling the height variations of the developed silver on the paper. It could watch a movie by walking its fingers along the film as it screamed by at high speed. There is no reason the fingers could not also be sensitive to light and temperature and other electromagnetic effects; indeed, the smallest are the right size to be "antennas" for light. The bush could form an eye by holding up a lens and putting a few million of its fingers in the focal plane behind it. It may even be able to get by without the lens by holding a bunch of its fingers in a carefully spaced diffraction pattern, thus forming a holographic lens.

In addition to having a sensing capability to match that of the world's current human population, our bush would have the ability to affect its environment at the same prodigious rate. A well-trained human, using precise and well-timed hand and body motions, each able to change direction at most a few times in a second, with a precision no better than a few percent of the total movement, could conceivably affect the world at a net rate of a thousand bits per second—a fast typist, for instance, produces less than one hundred bits per second of text. The potential data rate of a robot with one trillion fingers, each able to move a million times per second, is more than a quadrillion (10^{15}) times greater. Such high data rates imply huge coordination of enormous processing power, but imagine the possibilities. The bush robot could reach into a complicated piece of delicate mechanical equipment—or even a living organism—simultaneously sense the relative position of millions of parts, some possibly as small as molecules, and rearrange them for a near-instantaneous repair. In most cases the superior touch sense would totally substitute for vision, and the extreme dexterity would eliminate the need for special tools.

An astronomical amount of thinking would be required to control this wonderful machine. Much of it might be handled by what in animals are called *reflex arcs*, small bits of nervous system near the site being controlled. Each of the small branchlets could contain enough of a computer to control most of the routine activity; only exceptional situations would require intervention from larger computers nearer the stem. If the branches also contained their own power source (think rechargeable battery) and a way of communicating remotely (radio, or sound vibrations of a few thousand synchronized cilia, would do), the bush could break into a coordinated swarm of smaller bushes. The smaller the individual bush, the less intelligent and less powerful it would be. It would be preprogrammed, and charged up, by its home stem to perform some function and then return as soon as possible to report and receive new instructions.

Small size would frequently be an advantage: a smaller robot can squeeze into smaller spaces. A tiny machine has a greater surface-area-to-weight ratio: while a large bush could walk securely along the floor, using its branches as so many nimble toes, a smaller machine should be able to walk on ceilings like a fly, with the tiny cilia holding onto microscopic cracks in the paint or sticking by molecular adhesion. Bushes could burrow by loosening particles of dirt and passing them backward, and swim efficiently by assuming a tight, streamlined shape, with the cilia forming a skin that pumps fluid to propel and also responds to the flow to prevent turbulence. Extremely small machines will have so much surface area for their weight that they can fly like insects, beating their cilia in patterns optimal for moving air.

The contribution of the bigger branches to the power and intelligence of the smaller ones can be visualized as a kind of reverse pyramid scheme. Each level (counting all the twigs at that level) of our robot has twice the volume of the next smaller level and thus room for twice the power supply and twice the computer. Two levels down, the ratio is four to one, and at three levels it is eight. If the control and power for each level are piped up from the branches three steps closer to the stem, the small branches can be four times as vigorous as otherwise. Only the stem and the first few branches radiating from it would be shortchanged. Since most of their power and attention is directed higher up in the tree, they may be incapable of much motion and may be relegated to providing a stable framework if they passively lock their joints.

A big question is how the control programs for such a beast would work. In the extreme case one could imagine a program that would try to determine the combination of actions of each individual joint that would best accomplish the desired task. This is almost certainly an example of a colossal NP (nondeterministic polynomial) problem that can be optimally solved only by essentially examining every possible combination of motions and picking the best (see Appendix 3). Such solutions are intractable these days even for simple manipulators that have only a handful (!) of fingers. Though computers will be vastly more powerful in the future, the problem posed by a system with a large number of fingers is much, much bigger still. From time to time an especially clever strategy for coordinating thousands or millions of fingers to accomplish a particular task may be discovered, and collections of clever strategies will be passed between individuals— the manual skills of the superintelligent era. But while humans teach skills as simple as how to tie a shoelace, the lessons of superintelligent machines may be more comparable to instructions for assembling an airliner. Finding the best possible solution for routine problems will usually be out of the question, but finding a good enough one may not be too hard. I imagine a divide-and-conquer strategy, where the stem considers the overall problem and generates plausible subtasks for each of the four subtrees immediately connected to itself. These further subdivide the problem and pass those fragments on, and so on. The smallest branches would receive simple commands like move to a certain position, or move until resistance is met. A command from the stem might be something like: *North bush—stay on left side of plane A, and right side of plane B, and apply net force vector V to object; East bush—stay on right of A and B, and resist any motion more than 10 cm from the axis; South bush—right of A, left of B, apply force negative V; West bush—left of A and B, and resist.* If a subproblem, as passed to a small bush, cannot be solved, a complaint would be sent back to the originating branch, which would then go back to the drawing board to try something else.

A bush robot would be a marvel of surrealism to behold. Despite its structural resemblance to many living things, it would be unlike anything yet seen on earth. Its great intelligence, superb coordination, astronomical speed, and enormous sensitivity to its environment would enable it to constantly do something surprising, at the same time maintaining a perpetual gracefulness. Two-legged animals have three or four effective gaits; four-legged animals have a few more. Two-handed humans have two or three ways to hold an object. A trillion-limbed device, with a brain to match, is an entirely different order of being. Add to this the ability to fragment into a cloud of coordinated tiny fliers, and the laws of physics will seem to melt in the face of intention and will. As with no magician that ever was, impossible things will simply *happen* around a robot bush. Imagine inhabiting such a body.

Transmigration

Some of us humans have quite egocentric world views. We anticipate the discovery, within our lifetimes, of methods to extend human life, and we look forward to a few eons of exploring the universe. The thought of being grandly upstaged in this by our artificial progeny is disappointing. Long life loses much of its point if we are fated to spend it staring stupidly at our ultra-intelligent machines as they try to describe their ever more spectacular discoveries in baby-talk that we can understand. We want to become full, unfettered players in this new superintelligent game. What are the possibilities for doing that?

Genetic engineering may seem an easy option. Successive generations of human beings could be designed by mathematics, computer simulations, and experimentation, like airplanes, computers, and robots are now. They could have better brains and improved metabolisms that would allow them to live comfortably in space. But, presumably, they would still be made of protein, and their brains would be made of neurons. Away from earth, protein is not an ideal material. It is stable only in a narrow temperature and pressure range, is very sensitive to radiation, and rules out many construction techniques and components. And it is unlikely that neurons, which can now switch less than a thousand times per second, will ever be boosted to the billions-per-second speed of even today's computer components. Before long, conventional technologies, miniaturized down to the atomic scale, and biotechnology, its molecular interactions understood in detailed mechanical terms, will have merged into a seamless array of techniques encompassing all materials, sizes, and complexities. Robots will then be made of a mix of fabulous substances, including, where appropriate, living biological materials. At that time a genetically engineered superhuman would be just a second-rate kind of robot, designed under the handicap that its construction can only be by DNA-guided protein synthesis. Only in the eyes of human chauvinists would it have an advantage—because it retains more of the original human limitations than other robots.

Robots, first or second rate, leave our question unanswered. Is there any chance that we—you and I, personally—can fully share in the magical world to come? This would call for a process that endows an individual with all the advantages of the machines, without loss of personal identity. Many people today are alive because of a growing arsenal of artificial organs and other body parts. In time, especially as robotic techniques improve, such replacement parts will be better than any originals. So what about replacing everything, that is, transplanting a human brain into a specially designed robot body? Unfortunately, while this solution might overcome most of our physical limitations, it would leave untouched our biggest handicap, the limited and fixed intelligence of the human brain. This transplant scenario gets our brain out of our body. Is there a way to get our mind out of our brain?

You've just been wheeled into the operating room. A robot brain surgeon is in attendance. By your side is a computer waiting to become a human equivalent, lacking only a program to run. Your skull, but not your brain, is anesthetized. You are fully conscious. The robot surgeon opens your brain case and places a hand on the brain's surface. This unusual hand bristles with microscopic machinery, and a cable connects it to the mobile computer at your side. Instruments in the hand scan the first few millimeters of brain surface. High-resolution magnetic resonance measurements build a three-dimensional chemical map, while arrays of magnetic and electric antennas collect signals that are rapidly unraveled to reveal, moment to moment, the pulses flashing among the neurons. These measurements, added to a comprehensive understanding of human neural architecture, allow the surgeon to write a program that models the behavior of the uppermost layer of the scanned brain tissue. This program is installed in a small portion of the waiting computer and activated. Measurements from the hand provide it with copies of the inputs that the original tissue is receiving. You and the surgeon check the accuracy of the simulation by comparing the signals it produces with the corresponding original ones. They flash by very fast, but any discrepancies are highlighted on a display screen. The surgeon fine-tunes the simulation until the correspondence is nearly perfect.

To further assure you of the simulation's correctness, you are given a pushbutton that allows you to momentarily "test drive" the simulation, to compare it with the functioning of the original tissue. When you press it, arrays of electrodes in the surgeon's hand are activated. By precise injections of current and electromagnetic pulses, the electrodes can override the normal signaling activity of nearby neurons. They are programmed to inject the output of the simulation into those places where the simulated tissue signals other sites. As long as you press the button, a small part of your nervous system is being replaced by a computer simulation of itself. You press the button, release it, and press it again. You should experience no difference. As soon as you are satisfied, the simulation connection is established permanently. The brain tissue is now impotent— it receives inputs and reacts as before but its output is ignored. Microscopic manipulators on the hand's surface excise the cells in this superfluous tissue and pass them to an aspirator, where they are drawn away.

The surgeon's hand sinks a fraction of a millimeter deeper into your brain, instantly compensating its measurements and signals for the changed position. The process is repeated for the next layer, and soon a second simulation resides in the computer, communicating with the first and with the remaining original brain tissue. Layer after layer the brain is simulated, then excavated. Eventually your skull is empty, and the surgeon's hand rests deep in your brainstem. Though you have not lost consciousness, or even your train of thought, your mind has been removed from the brain and transferred to a machine. In a final, disorienting step the surgeon lifts out his hand. Your suddenly abandoned body goes into spasms and dies. For a moment you experience only quiet and dark. Then, once again, you can open your eyes. Your perspective has shifted. The computer simulation has been disconnected from the cable leading to the surgeon's hand and reconnected to a shiny new body of the style, color, and material of your choice. Your metamorphosis is complete.

For the squeamish, there are other ways to work the transfer of human mind to machine. A high-resolution brain scan could, in one fell swoop and without surgery, make a new you "While-U-Wait." If even the last technique is too invasive for you, imagine a more psychological approach. A kind of portable computer (perhaps worn like magic glasses) is programmed with the universals of human mentality, your genetic makeup, and whatever details of your life are conveniently available. It carries a program that makes it an excellent mimic. You carry this computer with you through the prime of your life; it diligently listens and watches; perhaps it monitors your brain and learns to anticipate your every move and response. Soon it can fool your friends on the phone with its convincing imitation of you.

When you die, this program is installed in a mechanical body that then smoothly and seamlessly takes over your life and responsibilities.

If you happen to be a vertebrate, there is another option that combines the sales features of all the methods described above. The vertebrate brain has two hemispheres connected by several large bundles of nerve fibers. The largest is called the *corpus callosum*. In the 1960s, guided by animal experiments, researchers in California successfully treated patients with intractable types of epilepsy by severing their corpora callosa. (Medical robots in the future will not use Latin!) Amazingly, this procedure appeared at first to have no side effects on the patients. The corpus callosum, with 200 million nerve fibers, is the brain's most massive long-distance interconnect. It is far thicker than the optic nerve or the spinal cord. Cut the optic nerve and the victim is utterly blind; sever the spinal cord and the body becomes limp and numb. But slice the huge cable between the hemispheres and nothing bad happens. Well, almost nothing. If the name of an object, like "brush," is flashed in the right half of the visual field of view of a "split-brain" person, the person is unable to select the object from among others with the left hand but has no difficulty making the choice with the right hand. Sometimes in the left-handed version of the task, the right hand—apparently in exasperation—reaches over to guide the left to the proper location!

Neuroanatomy suggests some of the explanation. The nerves directing the muscles of the left side of the body, as well as those portions of the optic nerve viewing the left half of the visual scene, are connected only to the right side of the brain. Conversely, the left side of the brain controls the right side of the body and sees the right half of the scene, as illustrated in the figure on page 113. Normally the two brain halves work in an intimate partnership, and information discovered by one is rapidly available to the other through the agency of the corpus callosum. In a split-brain person, this information flow is broken. The two brain halves must discover things independently. The left hand knows not what the right is doing. The two halves still seem to be aware of each other's emotions, however, from information apparently relayed through intact connecting nerves in the brainstem.

Roger Sperry of the California Institute of Technology, who received a Nobel Prize in 1981 for his discoveries on the function of the corpus callosum, found that in split-brain subjects each brain half seems to host an independent, fully conscious, intelligent human personality. In intact brains some of the corpus callosum fibers are known to handle basic functions such as recombining the halves of the visual fields of the eyes, but others must communicate higher mental concepts between the hemispheres. There is every reason to believe the corpus callosum provides a neatly organized and very wide window into the mental activities of both hemispheres. Suppose in the future, when the function of the brain is sufficiently understood, your corpus callosum is severed and cables leading to an external computer are connected to the severed ends. The computer is programmed at first to pass the traffic between the two hemispheres and to eavesdrop on it. From what it learns by eavesdropping, it constructs a model of your mental activities. After a while it begins to insert its own messages into the flow, gradually insinuating itself into your thinking, endowing you with new knowledge and new skills. In time, as your original brain faded away with age, the computer would smoothly assume the lost functions. Ultimately your brain would die, and your mind would find itself entirely in the computer. Perhaps, with advances in high-resolution scanning, it will be possible to achieve this effect without messy surgery: you might simply wear some kind of helmet or headband that monitored and altered the interhemispheric traffic with carefully controlled electromagnetic fields.

Many Changes

Whatever style of mind transfer you choose, as the process is completed many of your old limitations melt away. Your computer has a control labeled "speed." It had been set at "slow," to keep the simulations synchronized with the old brain, but now you change it to "fast," allowing you to communicate, react, and think a thousand times faster. The entire program can be copied into similar machines, resulting in two or more thinking, feeling versions of you. You may choose to move your mind from one computer to another that is more technically advanced or better suited to a new environment. The program can also be copied to a future equivalent of magnetic tape. Then, if the machine you inhabit is fatally clobbered, the tape can be read into a blank computer, resulting in another you minus your experiences since the copy. With enough widely dispersed copies, your permanent death would be highly unlikely.

The Corpus Callosum

The left half of the brain controls the right side of the body, and vice-versa. The left half usually specializes in language and calculation, while the right half is good at spatial reasoning. The brain halves normally communicate through the corpus callosum, but they can continue to function as separate individuals if it is severed.

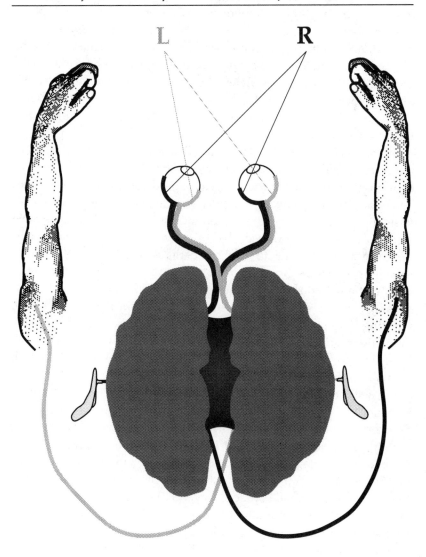

As a computer program, your mind can travel over information channels, for instance encoded as a laser message beamed between planets. If you found life on a neutron star and wished to make a field trip, you might devise a way to build a robot there of neutron stuff, then transmit your mind to it. Since nuclear reactions are about a million times quicker than chemical ones, the neutron-you might be able to think a million times faster. You would explore, acquire new experiences and memories, and then beam your mind back home. Your original body could be kept dormant during the trip and reactivated with the new memories when the return message arrived—perhaps a minute later but with a subjective year's worth of experiences. Alternatively, the original could be kept active. Then there would be two separate versions of you, with different memories for the trip interval.

Your new abilities will dictate changes in your personality. Many of the changes will result from your own deliberate tinkerings with your own program. Having turned up your speed control a thousandfold, you notice that you now have hours (subjectively speaking) to respond to situations that previously required instant reactions. You have time, during the fall of a dropped object, to research the advantages and disadvantages of trying to catch it, perhaps to solve its differential equations of motion. You will have time to read and ponder an entire on-line etiquette book when you find yourself in an awkward social situation. Faced with a broken machine, you will have time, before touching it, to learn its theory of operation and to consider, in detail, the various things that may be wrong with it. In general, you will have time to undertake what would today count as major research efforts to solve trivial everyday problems.

You will have the time, but will you have the patience? Or will a thousandfold mental speedup simply incapacitate you with boredom? Boredom is a mental mechanism that keeps you from wasting your time in profitless activity, but if it acts too soon or too aggressively it limits your attention span, and thus your intelligence. One of the first changes you will want to make in your own program is to retard the onset of boredom beyond the range found today in even the most extreme intellectuals. Having done that, you will find yourself comfortably working on long problems with sidetracks upon sidetracks. In fact, your thoughts will routinely become so involved that they will call for an increase in your short-term memory. Your long-term memory also will have to be boosted, since a month's worth of events will occupy a subjective span of a century! These are but the first of many changes.

I have already mentioned the possibility of making copies of oneself, with each copy undergoing its own adventures. It should be possible to merge memories from disparate copies into a single one. To avoid confusion, memories of events would indicate in which body they happened, just as our memories today often have a context that establishes a time and place for the remembered event. Merging should be possible not only between two versions of the same individual but also between different persons. Selective mergings, involving some of another person's memories and not others, would be a superior form of communication, in which recollections, skills, attitudes, and personalities can be rapidly and effectively shared. Your new body will be able to carry more memories than your original biological one, but the accelerated information explosion will ensure the impossibility of lugging around all of civilization's knowledge. You will have to pick and choose what your mind contains at any one time. There will often be knowledge and skills available from others superior to your own, and the incentive to substitute those talents for yours will be overwhelming. In the long run you will remember mostly other people's experiences, while memories you originated will be incorporated into other minds. Concepts of life, death, and identity will lose their present meaning as your mental fragments and those of others are combined, shuffled, and recombined into temporary associations, sometimes large, sometimes small, sometimes long isolated and highly individual, at other times ephemeral, mere ripples on the rapids of civilization's torrent of knowledge. There are foretastes of this kind of fluidity around us. Culturally, individual humans acquire new skills and attitudes from others throughout life. Genetically, in sexual populations each individual organism is a temporary bundling of genes that are combined and recombined in different arrangements every generation.

Mind transferral need not be limited to human beings. Earth has other species with large brains, from dolphins, whose nervous systems are as large and complex as our own, to elephants, other whales, and perhaps giant squid, whose brains may range up to twenty times as big as ours. Just what kind of minds and cultures these animals possess is still a matter of controversy, but their evolutionary history is as long as ours, and there is surely much unique and hard-won information encoded genetically in their brain structures and their memories. The brain-to-computer transferral methods that work for humans should work as well for these large-brained animals, allowing their thoughts, skills, and motivations to be woven into our cultural tapestry. Slightly different methods, that focus more on genetics and physical makeup than on mental life, should allow the information contained in other living things with small or no nervous systems to be popped into the data banks. The simplest organisms might contribute little more than the information in their DNA. In this way our future selves will be able to benefit from and build on what the earth's biosphere has learned during its multibillion-year history. And this knowledge may be more secure if it is preserved in databanks spreading through the universe. In the present scheme of things, on our small and fragile earth, genes and ideas are often lost when the conditions that gave rise to them change.

Our speculation ends in a supercivilization, the synthesis of all solar-system life, constantly improving and extending itself, spreading outward from the sun, converting nonlife into mind. Just possibly there are other such bubbles expanding from elsewhere. What happens if we meet one? A negotiated merger is a possibility, requiring only a translation scheme between the memory representations. This process, possibly occurring now elsewhere, might convert the entire universe into an extended thinking entity, a prelude to even greater things.

What Am I?

The idea that a human mind can be transferred to a new body sometimes meets the following strong objection from people who do not dispute the theoretical possibility: "Regardless of how the copying is done, the end result will be a new person. If it is I who am being copied, the copy, though it may think of itself as me, is simply a self-deluded impostor. If the copying process destroys the original, then I have been killed. That the copy may then have a great time exploring the universe using my name and my skills is no comfort to my mortal remains."

This point of view, which I will call the *body-identity position*, makes life extension by duplication considerably less personally interesting. I believe the objection can be overcome by acceptance of an alternative position which I will call *pattern-identity*. Body-identity assumes that a person is defined by the stuff of which a human body is made. Only by maintaining continuity of body stuff can we preserve an individual person. Pattern-identity, conversely, defines the essence of a person, say myself, as the *pattern* and the *process* going on in my head and body, not the machinery supporting that process. If the process is preserved, I am preserved. The rest is mere jelly.

The body-identity position, I think, is based on a mistaken intuition about the nature of living things. In a subtle way, the preservation of pattern and loss of substance is a normal part of everyday life. As we humans eat and excrete, old cells within our bodies die, break up, and are expelled and replaced by copies made of fresh materials. Most of our body is renewed this way every few years. The few body components such as nerve cells that tend to be more static nevertheless have metabolisms that cause their inner parts to be replaced, bit by bit. Every atom present within us at birth is likely to have been replaced half way through our life. Only our pattern, and only some of it at that, stays with us until our death.

Let me explore some of the consequences of the pattern-identity position. Matter transmitters have appeared often in the science fiction literature, at least since the invention of facsimile machines in the late 1800s. I raise the idea here only as a thought experiment, to simplify some of the issues in my mind-transfer proposal. A facsimile transmitter scans a photograph line by line with a light-sensitive photocell and produces an electric current that varies with the brightness of the scanned point in the picture. The varying electric current is transmitted over wires to a remote location, where it controls the brightness of a light bulb in a facsimile receiver. The receiver scans the bulb over photosensitive paper in the same pattern as the transmitter. When this paper is developed, a duplicate of the original photograph is obtained. This device was a boon to newspapers, who were able to get illustrations from remote parts of the country almost instantly, instead of waiting for a train to deliver photographic plates.

If pictures, why not solid objects? A *matter transmitter* might scan an object and identify its atoms or molecules one at a time, perhaps removing them in the process. The identity of the atoms would be transmitted to a receiver, where a duplicate of the original object would be assembled in the same order from a local supply of atoms.

The technical problems are mind-boggling, but the principle is simple to grasp, as millions of devotees of "Star Trek" will attest.

If solid objects, why not a person? Just stick him in the transmitter, turn on the scan, and greet him when he walks from the receiver. But is he really the same person? If the system works well, the duplicate will be indistinguishable from the original in any substantial way. Yet, suppose that you fail to turn on the receiver during the transmission process. The transmitter will scan and disassemble the victim and send an unheard message to the inoperative receiver. The original person will be dead. Doesn't the process, in fact, kill the original person whether or not there is an active receiver? Isn't the duplicate just that, merely a copy? Or suppose that two receivers respond to the message from one transmitter. Which, if either, of the two duplicates is the real original?

The body-identity position on this question is clear. A matter transmitter is an elaborate execution device that kills you and substitutes a clever impostor in your place. The pattern-identity position offers a different perspective. Suppose that I step into the transmission chamber. The transmitter scans and disassembles my jellylike body, but my pattern (me!) moves continuously from the dissolving jelly, through the transmitting beam, and ends up in other jelly at the destination. At no instant was the pattern (I) ever destroyed. But what about the question of duplicates? Suppose that the matter transmitter is connected to two receivers instead of one. After the transfer there will be a copy of me in each one. Surely at least one of them is a mere copy: they cannot both be me, right? *Wrong!*

Rooted in all our past experience is the assumption that one person corresponds to one body. In light of the possibility of separating mind from matter and storing and transmitting it, this simple, natural, and obvious identification becomes confusing and misleading. Consider the message, "I am not jelly." As I type it, it goes from my brain into the keyboard of my computer, through myriads of electronic circuits, and over great amounts of wire. After countless adventures, the message shows up in bunches of books like the one you are holding. How many messages were there? I claim that it is most useful to think there is only one, despite its massive replication. If I repeat it here: "I am not jelly," there is still only one message. Only if I change it in a significant manner ("I am not peanut butter") do we have a second message. And the message is not destroyed until the last written version is lost and until it fades sufficiently in everybody's memory to be unreconstructable. The message is the information conveyed, not the medium on which it is encoded. The "pattern" that I claim is the real me has the same properties as this message. Making a momentary copy of my state, whether on tape or in another functional body, does not make two persons.

The "process" aspect is a little more complicated. At the instant that a "person message" is assembled, it is just another copy of the original. But if two copies are active, they will in time diverge and become two different people. Just how far this differentiation must proceed before society grants them unique identities is about as problematical as the questions "When does a fetus become a person?" or "When does an evolving species become a different species?" But if we wait zero time, then both copies are the same person—if we immediately destroy one, the person still exists in the other copy. All the deeds that that person might have done, and all the thoughts she might have thought, are still possible. If, instead, we allow both copies to live their separate lives for a year and then destroy one, we are the murderers of a unique human being. But if we wait just a short time before destroying one copy, then only a little unique information is lost.

This rationale might be a comfort if you were about to encounter danger but knew that a tape copy of you had been made recently. Should you die, an active copy made from the tape could resume your life. This copy would differ slightly from the version of you that died, in that it lacked the memories since the time of copy. But a small patch of amnesia is a trivial affair compared with the total loss of memory and function that results from death in the absence of a copy.

Intellectual acceptance that a secure and recent backup of you exists would not necessarily protect you from an extreme desire to preserve yourself if faced with imminent death, even in a worthwhile cause. Such feelings would be an evolutionary hangover from your one-copy past, no more in tune with reality than fear of flying is an appropriate response to present airline accident rates. Old instincts are not automatically erased when the rules of life are suddenly rewritten.

The pattern-identity position has clear dualistic implications—it allows the mind to be separated from the body. Though mind is entirely the consequence of interacting matter, the ability to copy it from one storage medium to another would give it an independence and an identity apart from the machinery that runs the program. The dualism will be especially apparent if we consider some of the different possibilities for encoding.

Some supercomputers have myriads of individual computers interconnected by a network that allows free flow of information among them. An operating system for this arrangement might allow individual processes to migrate from one processor to another in mid-computation, in a kind of juggling act that permits more processes than there are processors. If a human mind is installed in a future machine of this variety, functions originally performed by particular parts of the brain might be encoded in particular processes. The juggling action would ensure that operations occurring in fixed areas in the original brain would move rapidly from place to place within the machine. If the computer is running other programs besides the mind simulation, then the simulation might find itself shuffled into entirely different sets of processors from moment to moment. The thinking process would be uninterrupted, even as its location and physical machinery changed continuously, because the pattern would be maintained.

The most direct way of performing complex operations such as occur in a human mind are often not the most efficient. A process that is described as a long sequence of steps in a program can sometimes be transformed mathematically into one that arrives at the same conclusion in far fewer operations. As a young boy, the famous mathematician Carl Friedrich Gauss was a school smart aleck. For a diversion, a teacher once set him the problem of adding up the all the numbers between 1 and 100. He returned with the correct answer in less than a minute. He had noticed that the hundred numbers could be grouped into fifty pairs, 1+100, 2+99, 3+98, 4+97, and so on, each pair adding up to 101. Fifty times 101 is 5,050, the answer, found without a lot of tedious addition. Similar speedups are possible in complex computer processes. So called *optimizing compilers* have repertoires of accelerating transformations, some quite radical, to streamline programs they translate. The key may be a total reorganization in the order of the computation and the representation of the data.

One powerful class of transformations takes an array of values and combines them in systematic ways to produce another array, so that each element of the new array is a number formed by a unique combination of all the elements of the old array. An operation on a single element in the new array can then often substitute for a whole host of operations on the original array, making enormous efficiencies possible. Analogous transformations in time also work: a sequence of operations can be changed into an equivalent one, where each new step does a tiny fraction of the work of every one of the original steps. The localized is diffused, and the diffuse is localized. A program can quickly be altered beyond recognition by a few mathematical rewrites of this power. Run on a multiprocessor, single events in the original formulation may appear in the transformed program only as correlations between events in remote machines at remote times. Yet in a mathematical sense the transformed computation is exactly the same as the original one.

If we were thus to transform a program that simulates a person, the person would remain intact: his mind is the abstract mathematical property that is shared by the old and the transformed programs; it does not depend on the particular form of its program. Mind, as I have defined it, is thus not only not tied to a particular body, it is not even bound to a particular pattern. It can be represented by any one of an infinite class of patterns that are equivalent only in a certain, very abstract way. (This observation tempts me into a brief philosophical extrapolation in Appendix 3.)

Immortality of the type I have just described is only a temporary defense against the wanton loss of knowledge and function that is the worst aspect of personal death. In the long run, our survival will require changes that are not of our own choosing. Parts of us will have to be discarded and replaced by new parts to keep in step with changing conditions and evolving competitors. Surviving means playing in a kind of cosmic Olympics, with each year bringing new events and escalated standards in old events. Though we are immortals, we must die bit by bit if we are to succeed in the the qualifying event— continued survival. In time, each of us will be a completely changed being, shaped more by external challenges than by our own desires. Our present memories and interests, having lost their relevance, will at best end up in a dusty archive, perhaps to be consulted once in a long while by a historian. Personal death as we know it differs from this inevitability only in its relative abruptness. Viewed this way, personal immortality by mind transplant is a technique whose primary benefit is to temporarily coddle the sensibility and sentimentality of individual humans. It seems to me that our civilization will evolve in the same direction whether or not we transplant our minds and join the robots.

The ancestral individual is always doomed as its heritage is nibbled away to meet short-term environmental challenges. Yet this evolutionary process, seen in a more positive light, means that we are already immortal, as we have been since the dawn of life. Our genes and our culture pass continuously from one generation to the next, subject only to incremental alterations to meet the constant demand for new world records in the cosmic games. And even within our personal life, who among us would wish to remain static, possessing for a lifetime the same knowledge, memories, thoughts, and skills we had as children? Human beings value change and growth, and our artificial descendants will share this value with us—their survival, like ours, will depend on it.

Awakening the Past

The ability to transplant minds will make it easy to bring to life anyone who has been carefully recorded on a storage medium. But what if some of the transcription has been lost? It should be possible to reconstruct many missing pieces from other information—the person's genetic code, for instance, or filmstrips of the person in life, samples of handwriting, medical records, memories of associates, and so on. Very effective sleuthing should be possible in a world of superintelligences with astronomical powers of observation and deduction. The pattern-identity position implies that a person reconstructed by inference would be just as real as one reconstituted from an intact tape. The only difference is that in the former case some of the person's pattern was temporarily diffused in the environment before being reassembled.

But what if no tape existed at all? Archaeologists today make plausible inferences about historical figures from scraps of old documents, pottery sherds, x-ray scans of mummified bodies, other known historical facts, general knowledge about human nature, and whatever else they can find. Creators of historical fiction use such data to construct detailed scenarios of how things might have happened. Superintelligent archaeologists armed with wonder-instruments (that might, for instance, make atomic-scale measurements of deeply buried objects) should be able to carry this process to a point where long-dead people can be reconstructed in near-perfect detail at any stage of their life.

Wholesale resurrection may be possible through the use of immense simulators. Powerful computers are used today to predict the course of the planets and spacecraft. The precise trajectory that took Voyager 2 past Jupiter, Saturn, Uranus, and their moons, and will soon take it by Neptune, was calculated by repeated simulations in which different starting times, directions, and velocities of the spacecraft were tried until a correct combination was found. More dramatically, if less accurately, modern weather programs simulate the action of the atmosphere over the entire globe. New aircraft designs, nuclear explosions, and an increasing number of other things are these days first tested in simulators. Such simulations give peeks into possible futures and thus confer the power to choose among them. Because the laws of physics are symmetrical in time, simulations can usually be run in reverse as well as forward and can be used to "predict" the past, perhaps guided by old measurements or archeological data. As with future predictions, any uncertainty in the initial measurements, or in the rule that evolves the initial state, will allow for several possible outcomes. If the simulation is detailed enough and is given all available information, then all of its "predictions" are valid: *Any of the possible pasts may have led to the present situation.*

This is a strange idea if you are accustomed to looking at the world in a strictly linear, deterministic way, but it parallels the uncertain world described by quantum mechanics. Now, imagine an immense simulator (I imagine it made out of a superdense neutron star) that can model the whole surface of the earth on an atomic scale and can run time forward and back and produce different plausible outcomes by making different random choices at key points in its calculation. Because of the great detail, this simulator models living things, including humans, in their full complexity. According to the pattern-identity position, such simulated people would be as real as you or me, though imprisoned in the simulator.

We could join them through a magic-glasses interface, which connects to a "puppet" deep inside the simulation and allows us to experience the puppet's sensory environment and to naturally control its actions. More radically, we could "download" our minds directly into a body in the simulation and "upload" back into the real world when our mission is accomplished. Alternatively, we could bring people out of the simulation by reversing the process—linking their minds to an outside robot body, or uploading them directly into it. In all cases we would have the opportunity to recreate the past and to interact with it in a real and direct fashion.

It might be fun to resurrect all the past inhabitants of the earth this way and to give them an opportunity to share with us in the (ephemeral) immortality of transplanted minds. Resurrecting one small planet should be child's play long before our civilization has colonized even its first galaxy.

5 | *Wildlife*

 T HE postbiological world will host a great range
of individuals constituted from libraries of accumulated knowledge.
In its early stages, as it evolves from the world we know, the scale and
function of these individuals will be approximately that of humans.
But this transitional stage will be just a launching point for a rapid
evolution in many novel directions, as each individual mutates by
dropping unneeded traits and adding new ones from the growing
data banks. A spectrum of scales will come to exist—from tiny, barely
intelligent configurations for tight spaces to star-spanning superminds
for big problems. The distinctions will not always be clear—a super-
mind might be composed of myriads of closely cooperating minor
intelligences, analogous in their interactions to an ant colony.

Superintelligence is not perfection—spectacular failures are certain.
For this reason, diversity is to be desired and expected. Independent
centers of activity will compete. Lines of development will meet
dead ends. Life, in fact, will proceed much as it does in the earth's
biosphere, only on a vaster, faster, and more diverse scale. And
though at first thought a leap out of our biological bodies might
seem to free us of the diseases of the flesh—alas, it is not so. As
with terrestrial life, freeloaders will lurk in the interstices of the
postbiological world, making uninvited livings at the expense of their
hosts. These pests will evolve from the same stuff that constitutes
polite society. Their effects often will be subtle enough to escape
detection, but some pests will be as large, powerful, and visible as
their hosts. In any case, interactions within a postbiological world
will share many characteristics of relations that shape the world we
know.

Trojan Horses, Time Bombs, and Viruses

If the world of artificial machinery has seemed disease-free so far, it is only because our machines have been too simple to support mechanical parasites. But computers have changed that, as they have changed so much else. Diseases have appeared in computer systems for at least two decades, but 1988 was the first big year for computer plagues, as almost every type of machine, large and small, was attacked by several "computer viruses" that were spread through computer networks and by promiscuous sharing of computer software. Most data diseases are deliberate constructions of playful or malicious programmers, but some evolved by accident.

Deceptive programs that have been called *Trojan horses* have appeared from time to time since the 1960s. Written by clever programmers as practical jokes, in response to a challenge, or for more shady reasons, Trojan horses masquerade as interesting or useful programs. But once activated, they may begin to secretly compromise restricted information, or erase the victim's disk files, or at least print scary messages. A dangerous variant of the Trojan horse delays its attack. Because the malign action is not immediately manifested, such *time bombs* in apparently harmless programs are likely to be copied much more widely and consequently do more damage when they eventually strike.

Until the late 1970s, software was distributed by hand in the form of punched paper tape, decks of punched cards, or magnetic tape. Distribution was limited, and it was often easy to trace the source of a program. The resulting accountability must have inhibited many infections at their source. I myself created a potentially nasty infection in 1968. The machine involved was a small one, an IBM 1130, that read its programs in the form of decks of punched cards fed in through a unit that could both read cards and punch them. In a burst of creativity lasting a few days, a friend and I created a single punched card that masqueraded as a "cold-start" card used for resuscitating the computer after a power shutdown or a crash. But instead of starting up the system, our card caused a duplicate of itself to be punched in every card following it. Used by an unsuspecting programmer, it would have destroyed a program deck and produced many copies of itself to create future havoc. I remember holding the innocent-looking card in my hand and contemplating its destructive power with awe.

With some reluctance, we decided to destroy all the copies of the card (after a few hours of testing, there were many copies of each of several variations of it around!). I like to think we did the right thing simply because of our good sense of values, but the probability that we would not remain anonymous must have weighed into the decision.

Programs sold for profit are easy to copy and often make their way to machines other than those of the original paying customers. This is certainly the case now in the era of personal computers, but even in the sixties and seventies it was perceived as a problem with large systems. Manufacturers of software for large computers and for micros have been known to insert time bombs to thwart unauthorized use of their wares. The more benign bombs merely prevent the copied program from working after a certain period. Perhaps the intent is to give the nonpaying customer time to become dependent on the program, so when it ceases to function he will be inclined to buy a legitimate version. More vindictive bombs have occasionally appeared. These usually delete files, but a few have exploited unusual properties of the hardware in order to physically damage the computer. Because paying customers have often been victimized by an accidentally triggered bomb, the general reaction to this approach to copyright violation is very negative, and the practice seems to have declined.

I am aware of several instances where time-sharing systems were assaulted by a more subtle variety of Trojan horse. The intent was not to vandalize or terrorize but to gain unauthorized access. In this case the program acts like a spy and uses its special location to uncover information, such as a victim's secret passwords, which it then records in a location accessible to its author. In some attacks the program mimics the computer operating system's "log-in" procedure, by which users gain access to the computer by typing their identification and a secret password. Another form of attack exploits the property of most operating systems that a running program acquires all of its user's file-access rights. A program whose cover is some useful service can then, surreptitiously, rummage through the victim's disk files for wanted information. In a successful attack, the victim remains oblivious of the breach.

In the late 1970s cheap personal computers created a new medium for software and software diseases. The spread of both was facilitated by computer bulletin boards, systems maintained by enthusiasts that allowed other computer owners to dial up and post messages and programs that could be accessed by anyone else who dialed in. Such facilities offered both anonymity and promiscuous sharing of data—and, as with the sexual revolution, a raft of opportunities for disease. By the early 1980s the newspapers began to report instances of random havoc in personal computers caused by programs downloaded from computer bulletin boards, programs that purported to be games, accounting software, or whatever.

The most virulent known form of software wildlife has been called a *virus*—a program fragment that, once inserted into a large program, acts to copy itself into other programs, just as a biological virus is a piece of genetic code that, once inserted into a cell, acts to copy itself into other cells. The analogy is a strong one, because today's million-bit computer programs have about the same information content as the genetic codes of bacteria, and the few thousand bits of a typical computer virus is a good match for the small genetic code of a biological virus. When a program containing a virus is invoked, the virus is momentarily activated. It looks through its unsuspecting victim's files for uninfected but accessible programs and inserts a copy of itself into one or more of them. These newly infected programs will repeat the process when they are themselves activated. Experimental tests of this idea conducted in the mid-1980s by Fred Cohen of the University of Southern California resulted in almost total infection of supposedly secure computer systems in less than a day. The infections easily spread from restricted users to ones with greater access to the files of others. System managers are exposed when, in keeping abreast of developments on their machines, they try out new programs announced by their users. Once a manager's programs are infected, the rest of a system quickly succumbs.

A virus that merely spreads is a minor nuisance, taking up a little storage space in its many copies and a little of the computer's time in its reproductive activities. One quiet little virus created in 1978, that infects only the operating system, has apparently spread to virtually every Apple II system disk in existence. But, as with a Trojan horse, a virus can carry with it instructions for espionage, sabotage, or theft. Several variants of a virus created as an act of terrorism were detected spreading among IBM personal computers in Israel early in 1988. Examination of the virus revealed that it was a time bomb programmed to erase files on Israel's 40th Independence Day. It was discovered early because it had a flaw—it could repeatedly infect the same program. In time infected disks became nearly full because their files were bloated with multiple copies of the virus. Viruses have naturally been a matter of special concern to people who are charged with government and commercial computer security. A cleverly engineered virus might infect the most secure parts of an international banking or national defense network, for example. A clever human embezzler or spy with only minor access to the system could create a software accomplice to liberate funds or secrets. Botched attempts of this kind have made the news from time to time. The successful assaults get no publicity.

Today's computer systems are like bodies with skins but no immune systems, or like walled cities without police. They can deflect some external attacks but are defenseless once an intruder has entered. Internal protection is possible, though no defense is perfect. One approach is simply to build more walls. The easiest mode of viral spread can be blocked by preventing one program from altering another one—but legitimate purposes would be inhibited as well. It would, for instance, no longer be possible to "patch" programs to correct newly discovered errors in them. A new facility for patching could be installed in the operating system, but it would then itself be a potential gateway for viruses.

Instead of erecting blockades, another approach is to actively hunt viruses down. And, in fact, a first generation of virus killers has appeared on the heels of the first generation of viruses. One kind is a program that examines other programs—it detects particular viruses by the telltale pattern of their instructions and removes any it finds. But if the system is kept running while the exorcism is under way, a virus may reproduce faster than the erasing program can stamp it out, since each infected program can be a site of spreading disease. One solution is to shut down everything in the system except the virus killer until all programs are clean. This works unless the purging program itself has become infected; moreover, absolutely every program in the system must be deloused, and a single trace of the virus from a backup tape or from an external source can reestablish the infection.

A more aggressive approach to combating a viral infection is with another virus. A viral predator, like its prey, insinuates itself from program to program. But instead of causing problems for the users, it deletes any copies of the offending virus it encounters. Since it can propagate itself to every program in the system that the original virus can reach, its many copies will eventually be at every possible site of infection of the prey virus, able to immediately quash it if it reappears. The killer virus could be left around indefinitely, conferring a permanent immunity against the prey virus, or it could be programmed to remove itself after a specified time or on receiving a signal, to save space and running time.

Computer Virus Blowup
A well-designed virus will normally copy itself into another program only once. But the test which detects that a program is already infected may be foiled by infection with a different virus that hides the first infection. Two (or more) viruses may thus repeatedly infect the same program by alternately foiling each other's detectors.

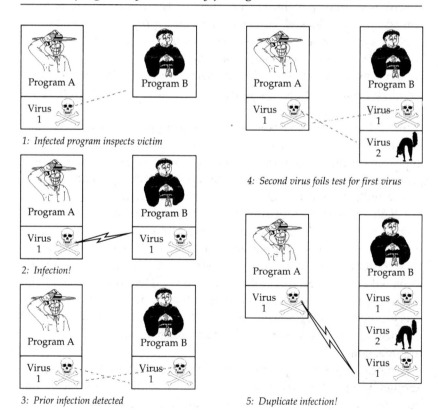

1: *Infected program inspects victim*

2: *Infection!*

3: *Prior infection detected*

4: *Second virus foils test for first virus*

5: *Duplicate infection!*

But no defense is perfect. A prey virus can be cosmetically altered so that it is no longer recognizable to a particular predator. Or the predator might mistake a portion of a legitimate program for a virus and erase it, thus breaking a working program. A viral predator might be altered by unintended interactions with other programs and be mutated into a pest that spreads too virulently or destroys the wrong pieces of code. Complicated situations can arise when there is more than one virus loose in a system. A well-designed virus, unlike the one unleashed in Israel, would check a potential host program to see if it had already been infected and refrain from duplicating the infection. But an intervening infection from a second virus might confuse this test and allow the first virus to insert a second copy of itself. Similarly, the new copy of the first virus might free the second virus to insert another copy, and so on, as in the figure on page 130, eventually bloating the program into uselessness. Even a viral predator might become entangled this way with a virus it failed to recognize.

Attempts by software manufacturers to protect their copyrights with time bombs have proven highly unpopular. But computer viruses may help their cause in a different way. Software downloaded from public bulletin boards or copied from friends is now always suspect—it may be a carrier of many contagions. But software purchased directly from a publisher can carry a guarantee of sterility backed by the publisher's reputation. Computer viruses may thus have the same effect on software piracy that the AIDS epidemic is having on sexual promiscuity.

Ghosts

A really intriguing vanishing act is possible because viruses can exist at different levels of abstraction—from patterns of wires connecting logic-gate assemblies, to patterns of bits controlling the opening and closing of paths between assemblies, to machine-language instructions commanding simple arithmetic operations, to letter strings of assembly programs abstractly representing machine language, to high-level languages expressing the goals of the programmer with little reference to the hardware details of how they are to be carried out. Wildlife can exist at all these levels and more.

Some viruses can move from one level of abstraction to another or become active at any level—or even at several levels. For example, consider a virus written in a high-level computer language. When a program containing the virus is compiled, it results in a machine-language program that searches for other programs in the high-level language and inserts the high-level version of itself into them. Such tampering in human-readable files is very easy to detect, but it offers the virus one great advantage: the ability to spread from one kind of computer and operating system to an entirely different one, since other compilers can translate it into other machine languages.

There are more subtle ways to achieve machine independence. A cleverly doctored compiler can insert the viral code into the programs it prepares. Modern compilers are often written in their own language and compiled by older versions of themselves. An archetype is the C compiler that constructs the popular Unix operating system. Partly because it is so easy to install on new machines, Unix has grown to be the most widespread operating system. Unix and C were invented in the early 1970s at Bell Labs by two young hackers, Ken Thompson and Dennis Ritchie, on their own initiative. AT&T, the owner of Bell Labs, took very little notice of it at first, until the system, improved by many programmers, had spread to thousands of machines worldwide. Finally, in the late 1970s, AT&T decided to sell Unix systems commercially. As part of the formalization of the product, AT&T assigned a programming team to extend and optimize it. During the course of careful scrutiny, someone noticed that the C compiler produced a large, completely unexplained block of machine-language program when it reached a certain point in compiling another C compiler. The discovery led to a number of closed-door meetings. In 1984 Ken Thompson, in a lecture on the occasion of his receiving the Turing Award (sort of a computer-science Oscar) shed light on the issue.

The programming team had uncovered a fiendishly clever virus, one designed to allow Ken Thompson to bypass security and to log into any Unix system running on any machine, including future machines yet to be invented. The mysterious block of code that appeared in new C compilers had two purposes. One was to cause a copy of itself to appear in any future C compilers that it itself compiled. The other was to insert a block of code that would respond to a password known to Ken Thompson, at a certain place in any log-in program that it compiled. The beauty of the scheme was that these blocks of code were reproduced not directly in machine language, which would work on only one type of machine, but by means of an ephemeral C version of themselves that was immediately translated by the compiler into the machine language of whatever machine it was compiling for. This self-reproducing C program had once existed in an actual computer file in Ken Thompson's machine but now had only a ghostly existence, reappearing momentarily deep in the innards of the computer whenever a C compiler executed the viral code and immediately vanishing again.

A ghost virus haunting a compiler is very hard to find even if one knows what to look for because a compiler is a huge program, and the virus appears only as an innocent-looking piece of code somewhere in the middle of a vast tract of legitimate machine language. The principle could be used to construct viruses with other purposes that could spread undetected between different kinds of machines. This example also hints at the possibility that even more subtle and elaborate creations are possible. Computers can bring mathematical abstractions to vigorous life, and there are no mathematical limits to the subtlety and deviousness. The fun has just begun.

Spontaneous Generation

So far I have confined this review of software parasites to those that are deliberately constructed, because these constitute the majority of the wildlife known to date. But such fabrications are limited by the imaginations of their human creators. Our increasingly complex systems are capable of creating their own surprises, and in times to come we can expect shockingly original gremlins to arise spontaneously in our intelligent machinery, the result of unexpected interactions or mutations of existing parts. A few stirrings have already been observed.

The ARPAnet, funded by the Department of Defense's Advanced Research Projects Agency, is a computer network created in the late 1960s to allow computers at facilities across the country to communicate with one another. The goal is to share scarce resources. Computers at individual sites are connected to the ARPAnet through special small computers once called Interface Message Processors, or IMPs. The IMP of one site is connected (through leased lines) only to those of a few nearby sites. Fancy software allows messages to cover greater distances by being handed rapidly from IMP to IMP. There are many possible indirect routes from point A to point B in the net. Depending on fluctuating traffic, sometimes one route is faster, sometimes another. To help make instantaneous routing decisions, each IMP maintains a table that records how long it takes recent messages to travel from the IMP through alternative routes to other sites. The table is updated with information obtained partially from the tables of neighboring IMPs. The network is monitored and maintained under contract by a company in Cambridge, Massachusetts. Operators at this site can stop, examine the contents, reload, and generally fiddle with any IMP on the net through special priority messages routed via the net itself. This generally works well, and even the most serious problems (such as large power outages affecting several IMPs) are handled smoothly.

In 1972 (and again in 1980, and probably other times as well) a plague hit the ARPAnet. The symptoms were that net traffic became hopelessly congested around a site in the Los Angeles area (in the 1980 incident the locus was Boston). Network control, suspecting some glitch in the program of the machine at the focus of the congestion, shut it down, reloaded its program, verified that it was working correctly, and reconnected it to the network. The problem persisted. Indeed, it seemed to be spreading outward from the original IMP. Shutting down and reloading larger numbers of IMPs did not fix things either; the congestion continued its spread and returned to the original sites as soon as they were reactivated. The network seemed to be haunted by a very persistent ghost. Many unsuccessful experiments later, order was finally restored by shutting down the entire network, clearing all the memories of all the IMPs, reloading their programs, and starting fresh—like sterilizing a whole planet with death rays, then seeding it with new life!

A subsequent analysis revealed what had happened. The focal IMP in Los Angeles had a memory error and developed an erroneous entry in its routing table. The table now indicated that messages sent via this IMP would experience a large *negative* delay. Next, adjacent IMPs calculated that it was more advantageous to send messages via this IMP than directly, since its negative delay more than made up for the extra hops. IMPs connected to *those* then decided it would be best to transmit via Los Angeles, and so on. The error in the initial IMP rapidly spread to the routing tables across the land. Wiping out the memories of a few IMPs did not clear the problem, because the erroneous numbers would spread back from IMPs that were still affected. Or infected. In fact, the network was inhabited by a spontaneously evolved, quite abstract, self-reproducing organism. This organism was formed by a simple, random mutation of a normal, sanctioned piece of data. It did not even involve a programming language.

The plague was easily spotted and eradicated because its effects were so devastating. If it had been more subtle in its action, it might have lived much longer. Among programs without masters there is a strong natural-selection criterion: reproduce but lie low. It is quite likely that many unsuspected organisms are already making a quiet living among the abstraction hierarchies in computer memories everywhere. Most will never be discovered. The plague also suggests a quick way in which a wild information organism can come into being, namely, by a mutation in an existing self-replicator. Since any datum in a computer is subject to duplication, this covers a lot of ground. A human-created computer virus, existing in many copies spread widely through different computer systems, would be a particularly fertile candidate for a liberating mutation. If the part of its code that caused it to make trouble were inactivated by a mutation, it would be less likely to draw attention to itself and thus would be more likely to reproduce indefinitely. The mutation might even make it unrecognizable, and thus safe, from a virus killer. Further mutations that eliminated unnecessary code, and thus reduced its size, would improve its chances even more. In time it might change in such a way as to fine-tune the frequency and kind of mutations it undergoes, making it more adaptable. Run-ins with other programs could endow it with major new pieces of code, and new capabilities. In time it might even acquire the ability to systematically copy and try out fragments of code from other programs and other viruses—the beginnings of computer-virus sex!

Such examples merely demonstrate the limits of our imagination. The most effective organisms would be much more subtly encoded and would escape detection entirely. From time to time one might expect one to surface because it developed a nasty side effect. That kind of mutation would generally prove fatal for the organism. As intelligence in programs progresses, we should also expect program fragments that can plan and act in a deliberate, calculating, and creative manner to enhance their survival. The data realm will host rats, coyotes, and master criminals as well as viruses and worms. Perhaps we will also be surprised by the equivalent of flowers, trees, and songbirds.

If these speculations are alarming, it may be comforting to remember that biological life thrives despite (or because of!) the relentless evolution of new parasites. Viruses insinuated themselves into the genetic machinery of cells long before their namesakes invaded computer programs. In biology also, information can be stored in various forms. Simple viruses inject DNA into cells, which then suicidally act on it to manufacture more viruses. The HTLV family of so-called retroviruses that cause AIDS and some kinds of leukemia, on the other hand, contain RNA that first must be transcribed to DNA, in a reversal of the normal synthesis. Some viruses use the time-bomb strategy, lying inactive in cells and thus evading immune system defenses until some event, possible signaling stress in their host and consequent reduced immunity, triggers their massive expression. Among the most effective parasites are sequences within DNA itself—called introns— that inhabit the genes and seem to do nothing at all except reproduce with their cell. Such long, repetitive sequences of DNA that played no apparent part in development had long been observed in the genetic codes of most organisms before it was suggested that their sole function might be to reproduce themselves. Richard Dawkins gives many more such examples in *The Selfish Gene* and *The Extended Phenotype*.

A Caveat for SETI

SETI, an acronym for the Search for Extra-Terrestrial Intelligence, is a field of study whose potential is so intellectually exciting that it proceeds steadily despite any hard evidence that its quarry exists. At its leading edge are impressive spectrum-analyzing receivers connected to radio telescopes that can tune in and examine millions of frequency channels at the same time. Systems able to do this and also look in thousands of distinct directions at once have already been proposed, all in an effort to find a needle in a haystack—an artificial message in a universe naturally noisy in radio frequencies.

But if we managed to receive and decode such a message, should we act on its instructions? The discussion of this question usually centers on the intent of the senders. They may be benign and, like the Peace Corps, be doing well by doing good. They may be traders trying to open new markets, to much the same effect, at least until it comes time to negotiate the price. They may simply be looking for pen pals. They may have dark designs on the rest of the universe and be seeking to inexpensively eliminate some of the more gullible competition. Or, their motives may be totally incomprehensible. Simply examining the message is not enough; it is not, in general, possible to deduce the effect of complicated instructions without actually carrying them out. A message with nasty intent would surely be disguised, by master deceivers, to look benign. In Fred Hoyle and John Elliot's classic novel *A for Andromeda* and also in Carl Sagan's *Contact*, an interstellar message contains plans for a mysterious machine of unknown purpose. In both books the characters decide, after some debate, to go ahead with construction despite the risks. In *Contact*, a major argument is that the origin of the message, the star Vega, is so close to our solar system that the senders could rapidly arrive here physically, should their intentions be malign. Building the machine would be unlikely to make us any worse off in the long run. If the message were benign, however, it represents an opportunity not to be missed.

This chapter's notion of an information parasite suggests greater caution, should SETI ever detect an artificial message. A rogue message from no one in particular to no one in particular (perhaps a corruption of some ancient legitimate interstellar telegram) could survive and thrive like a virus, using technological civilizations as hosts. It might be as simple as, "Now that you have received and decoded me, broadcast me in at least ten thousand directions with ten million watts of power. *Or else.*" It would be a cosmic chain letter and a cosmic joke, except to the message itself which, like any living creature, would be making a living by doing what it does. Since we cannot be sure the "or else" is not backed by real authors with a peculiar sense of right and wrong, we may decide to play safe and pass the message on as it requests. Perhaps we did not hear it very well; maybe it said a hundred million watts; maybe it mutated. Now envisage a universe populated by millions of such messages, evolving and competing for scarce, gullible civilizations.

The survivability of such a message could be enhanced if it carried real information. Perhaps it would contain blueprints for a machine that promises to benefit its hosts. It would be only fair if part of the machine's action was to rebroadcast copies of the message itself, or to demand new information from its hosts to be added to the message to make it more attractive to future recipients. Like bees carrying pollen for the sake of flowers in return for nectar for themselves, the technological host civilizations would have a symbiotic relationship with such messages, which might be criss-crossing the galaxy trading in useful ideas. But the analogy suggests darker possibilities. Some carnivorous plants attract bees with nectar, only to trap them. The message may promise a benefit, but when the machine is built it may show no self restraint and fiendishly co-opt all of its host's resources in its message sending, leaving behind a dead husk of a civilization. It is not too hard to imagine how such a virulent form of a free-living message might gradually evolve from more benign forms. A "reproduction effort parameter" in the message (too subtle for the victims to catch and alter) may get garbled in transmission, with the higher settings resulting in more aggressive and successful variants.

The *Fermi paradox* is an observation by the famous physicist Enrico Fermi, who created the first controlled atomic chain reaction under the auspices of the Manhattan Project, that if technological civilizations have even a slight probability of evolving, their presence should be visible throughout the universe. Our own history and prospects suggest that we will soon blossom into the universe ourselves, leaving it highly altered in our wake. In less than a million years we may have colonized the galaxy. Given the great age of the universe, a few civilizations that arose before us should have had plenty of time to alter many galaxies. The sky should be filled with the cosmic equivalent of roaring traffic and flashing neon signs. But instead we perceive a great silence.

There are several possible explanations. Evolutionary biologists make a plausible, though not watertight, argument which notes that at each stage of our evolution there were an immense number of evolutionary lines which did *not* head toward high technology, as compared with the single one that did. By this argument, we are the product of a sequence of very improbable accidents, a series unlikely to have been repeated in its entirety anywhere else. We may be the first and only technological civilization in the universe. But there are other explanations for the great silence. At the height of the cold war, a leading one was that high technology leads rapidly to self-destruction by nuclear holocaust or worse. But in every single case? Another possibility is that advanced civilizations inevitably evolve into forms that leave the physical universe untouched—perhaps they transmute into an invisible form or escape to somewhere more interesting. I discuss such a possibility in the next chapter.

A frightening explanation is that the universe is prowled by stealthy wolves that prey on fledgling technological races. The only civilizations that survive long would be ones that avoid detection by staying very quiet. But wouldn't the wolves be more technically advanced than their prey, and if so what could they gain from their raids? Our autonomous-message idea suggests an odd answer. The wolves may be simply helpless bits of data that, in the absence of civilizations, can only lie dormant in multimillion-year trips between galaxies or even inscribed on rocks. Only when a newly evolved, country bumpkin of a technological civilization stumbles and naively acts on one does its eons-old sophistication and ruthlessness, honed over the bodies of countless past victims, become apparent. Then it engineers a reproductive orgy that kills its host and propagates astronomical numbers of copies of itself into the universe, each capable only of waiting patiently for another victim to arise. It is a strategy already familiar to us on a small scale, for it is used by the viruses that plague biological organisms.

Pestilence as Positive

Is parasitism merely an unavoidable evil? If we could eliminate it— a highly unlikely prospect—should we? Perhaps not. A perfectly planned process is devoid of surprises; it is limited by the imagination of its designers. What new ideas and insights, otherwise unnoticed, might be *harvested* from freely evolving digital wildlife? Like the diverse genes of wild plants and animals, which are fuel for the advance of agriculture, the surprises in our machines sometimes point to profound truths, or at least to useful engineering tricks.

It has been argued that we biological beings owe our best features to the presence of diseases and other parasites in the world. It all has to do with sex. The earliest organisms reproduced asexually, repeatedly dividing into identical copies except when, from time to time, an individual cell became changed and passed on the mutation. In a well-functioning complex system such as a cell, a random change is extremely unlikely to be beneficial. So most mutations, if not immediately fatal, put their possessors at a disadvantage, and they vanish eventually in the press of competition for food and space. But once in a very long while a change for the better just happens to happen. The lucky owner of a beneficial mutation then has an advantage over its competing relatives, and over many generations its descendants will become a large fraction of the population. Because the odds of a beneficial mutation are so low, only when there are many copies of one beneficial mutation in existence does a second good mutation stand a reasonable chance of joining it. In an asexual species, each beneficial mutation has a refractory period before it can be compounded by another one.

But in a population where individuals can share genes sexually, two beneficial mutations that arise separately in different individuals can combine rapidly to form an offspring with both advantages. The effect is an acceleration of evolution. It is thus no accident that all higher organisms reproduce sexually (or had ancestors that did). That is how they got to be higher organisms so quickly. The asexual organisms, for the most part, are still swimming around as single cells or in small colonies. Acceleration of the evolutionary rate can be viewed as a long-term advantage of sexuality. In the short run, though, sex is a liability, because it increases the cost of reproduction. Instead of simply dividing whenever conditions seem right and producing a daughter that carries 100% of oneself, one must go to the trouble of finding a mate to produce an offspring that is only 50% true. Why, then, would sex ever arise? And if it did, why does it not disappear in a few generations under the onslaught of the more effective asexual reproducers?

Enter disease. In asexual reproduction, according to an evolutionary theory first developed by William D. Hamilton, each individual is an identical copy—a clone—of every other one. If a parasite evolves that can breach the defenses of one individual, then it can conquer every other. Like a wildfire, it can destroy a whole community in short order. In a sexual population, though, each individual is the result of a unique shuffle of genes taken from a large pool and is, in general, different from every other individual. A parasite that has the key to one lock finds that the next one is subtly different and thus harder to open. In a pest-filled world, the diverse sexual population does better than the homogeneous asexual community.

If disease made us sexy and sexiness made us smart, we can expect that digital wildlife will similarly make the data world more hardy, more diverse, and much more interesting.

Selfish Altruism

Some competition may be a good thing, but is the postbiological world fated to unmitigated cutthroat strife at every level of abstraction? Fortunately for the sake of organized existence, the answer appears to be a qualified no. In *The Evolution of Cooperation*, the political scientist Robert Axelrod notes that cooperation in the biological world can be observed in situations ranging from relations between large creatures and their microbial inhabitants to relations of human beings with one another. In a world where selfishness usually pays off, he asks, how could altruism between unrelated individuals ever arise?

To find an answer, Axelrod challenged game theorists, biologists, sociologists, political scientists, and hackers to submit computer programs which would compete in tournaments that modeled, in a highly abstract form, the typical costs and benefits of cooperation and its opposite, defection. Programs were paired in contests whose outcomes modeled the so-called *prisoner's dilemma* of game theory. This paradoxical situation was originally cast as a problem faced by two partners in crime who have been apprehended with insufficient evidence for conviction but who are given tempting inducements to inform on one another. In Axelrod's tournament, without knowing its opponent's choice, each competing program was given the choice of either cooperating or defecting. If both cooperated, each would receive the moderate "nice's reward." If both defected, each would receive the smaller "nasty's payoff." If one cooperated and the other defected, the defector would receive a huge "cheater's spoils," while the cooperator would receive nothing—the "sucker's payoff." So, if player B were to cooperate, player A could win the cheater's spoils by defecting, and only the lesser nice's reward by cooperating. If, on the other hand, player B defected, player A would at least get the nasty's payoff by defecting, instead of being left empty-handed with the sucker's payoff for cooperating. In other words, regardless of player B's choice, player A does better by defecting. So, obviously, to get the highest score, A should defect. The same reasoning applies to B. So both should defect, even though mutual cooperation would lead to a higher payoff! This is the crux of the prisoner's dilemma, and a feature of many interactions between selfish individuals that makes cooperation, whatever its theoretical benefits, seem very unlikely.

Imagine two selfish Martians crossing paths one fine sunny day while both are backpacking on the plains near Mars' great volcano, Olympus Mons. It turns out that one Martian has a small supply of batteries, while the other has some empty flashlights. Unfortunately, a few batteries happen to be dead, and some of the flashlights are burnt out—in each case the owner knows which units are defective, but there is no way for the other Martian to tell, because Martian electrical devices use superconductors that work only in the bitter cold of night. The Martians agree to exchange a flashlight for a battery, and then be on their way, never to meet again. This is a prisoner's

Selfish Martians

dilemma situation. Giving a good unit in the exchange counts as cooperation, while giving a defective unit is a defection. The Martians would benefit from mutual cooperation—they would then each have a working flashlight—but there is no incentive to give away a good battery or light that will be of use later, since good and bad units are indistinguishable by day. Each Martian gives the other a broken unit and leaves the meeting gloating over having made a shrewd deal. But as night falls, both Martians find themselves in the dark.

In Axelrod's tournament the players met repeatedly, so that each player could use its opponent's past behavior to shape its next move. The fifteen strategies that were submitted, whether simple or elaborate, fell into two categories—nice and nasty. Nice programs never defected first, giving nasty ones a potential temporary advantage. The simplest program was a nasty one, called *All D*, which always defected. The next simpler, submitted by Anatol Rapoport, a psychologist and game theorist at the University of Toronto, was *Tit for Tat*, which cooperated on the first encounter with any player and then on subsequent moves reciprocated the other player's previous move. A control entrant was *Random*, which randomly defected or cooperated, with equal probability, on each move.

To Axelrod's surprise, the nice program *Tit for Tat* won the first round, as well as later rounds with larger numbers of players. The result is surprising, because *Tit for Tat* defects only once for each defection against it, and so can gain no more from a sequence of interactions than its opponent and is likely to get less against a nasty player, because it will be cheated on the first move. Yet its overall score was highest, and the nice programs as a group performed much better than the nasty programs. The explanation, in game-theory terms, centers on the fact that the interaction was not *zero sum*. In a zero-sum game, a gain for one player is an equal loss for the other. In the prisoner's dilemma, however, both players can do better by cooperating. Nice programs interacting with others of their kind always benefit from the nice's reward. Nasty–nasty interactions conversely result in only the nasty's payoffs. Most of the nice programs eventually refused to cooperate with nastys, so though the nasty programs gained an initial advantage by cheating, they suffered eventually by forfeiting the rewards of cooperation. Slightly nasty programs that tried to gain a small advantage by occasionally defecting often initiated mutual retaliation cycles, giving them a net large disadvantage.

Axelrod's conclusion is that cooperation pays when the likelihood of future interactions with identifiable individuals is reasonably high. If the game is likely to end soon, however, cheating is the more successful strategy, since there will be little opportunity for retaliation. The theory seems to apply over a broad range of circumstances; the participants and the payoffs on the two sides of the game need not be commensurate, so long as on each side the payoffs for cheating, mutual cooperation, mutual defection, and being suckered are in descending order.

Axelrod provides several especially intriguing speculations on how this theory may be applied to the natural world. Large animals are inhabited by entire ecologies of microscopic fauna, most of which live quite peacefully with their hosts. Occasional infectious and fatal flareups of endogenous microorganisms show that this blissful state is not the only possibility. In fact, the relationship has the character of the prisoner's dilemma. The relation of an animal to its microscopic cohabitants is a selfish one—both animals and bacterial colonies are designed primarily to ensure their own survival. Though neither the microorganisms nor their host know each other in a personal way, the identity of each is assured by the constancy of the cohabitation. The microfauna can "defect" by overbreeding or releasing toxins and injuring or killing their host, but they are likely to be met by a defection in turn when the host fails to maintain a comfortable environment. The converse obtains when the host "tames" its parasites by rewarding good citizenship. In this way an initially hostile relationship can settle down to something more mutually beneficial. But the prisoner's dilemma remains, and if future interactions become unimportant, defection may again become advantageous.

For example, under trauma, like a large perforation of the gut wall, some of the normally friendly bacteria in an animal's gut change character and become seriously, even fatally, infectious. Axelrod and his colleague Hamilton speculate that this is an example of defection when future interactions are unimportant. The trauma is a signal to the bacteria that the game may be about to end, causing them to break the cooperative relationship to gain a last-minute advantage. By reproducing massively at the expense of their host, they can perhaps broadcast enough spores to find residences elsewhere. Presumably their ancestors survived the demise of other hosts by this strategy.

Axelrod's insights into cooperation, though no doubt just the tip of the iceberg, suggest that the chaos threatened earlier in this chapter will sort itself out most of the time. Parasitic relations at whatever level of abstraction will often become less destructive and may even turn symbiotic, since both partners thus do better. Harmony so achieved is not guaranteed. In certain circumstances defection can carry an advantage, and the truce can collapse. The net effect on future intelligences and systems will then be greater unpredictability. Information viruses that arise within a system and then vanish from view after achieving a peaceful, cooperative lifestyle will nevertheless modify the overall behavior of the system in subtle ways. Mature systems may become more a product of tamed pests than of their original design. Our best-laid plans are thus foiled, but conversely our descendants are spared the consequences of the limits in our vision. Our intelligence can control the future only imperfectly, and only in the near term.

Leaving the far future to the fates, does superintelligence help at all in making the world a nicer place in the near term? Axelrod observes that cooperation can arise even in populations of defectors. It does not depend on any intelligence in the participants—simple natural selection is a perfectly adequate driver. It does require that a certain minimal number of cooperators appear simultaneously to benefit from one another's niceness. Getting a critical mass of cooperators may take a long time. Intelligence can help because it allows individuals to anticipate the long-term advantages of initiating pleasantries. The long memories of the long-lived individuals who will inhabit the postbiological world are also likely to enhance the advantages of being nice, since no interaction is likely to be the last. Beyond the scope of Axelrod's tournament, intelligence allows one individual to learn about another's character by observing its interaction with third parties. The computer scientist Douglas Hofstadter goes so far as to imagine that in games between superintelligences, cooperation will be the rule *even when no future interactions are expected*. Each player will reason that all the players, being rational, would make the same decision as himself. Thus a defection will be met by defection and cooperation by cooperation. Maybe so, but there is always the possibility that a cooperator will be suckered by a devious opponent who, for whatever reason, foresees no future interactions.

Despite the likelihood of cooperative behavior on the large scale, and in the long run at every level, there will be occasional appearances of nasty little parasites. Permanent structures analogous to immune systems and police forces will undoubtedly be a part of large organisms. I expect a future world friendly overall, but with pockets of fruitful chaos at most levels.

6 | *Breakout*

OUR descendants may have a good time for a long while, developing their minds, exploring the universe, mastering space and time, the very large and the very small. But doesn't the second law of thermodynamics assure that the fun will eventually end? The study of the theory of steam engines led to one of the greatest scientific shocks of the nineteenth century—the realization that the universe is running down. Hot things get cooler and cold things get warmer, and energy once available for engines great and small will become inaccessibly lost in a uniform jumble of molecular motion. In time the entire universe will become a homogeneous stew with no concentrations of matter or energy to form or power any kind of machinery, intelligent or otherwise. This regressive idea of a *heat death* greatly disturbed Victorian minds attuned to steady progress in both society and Darwinian nature.

Fortunately for my own hopes for the future, twentieth-century physics and cosmology have loosened the hold of the second law. Instead of a closed, static universe, we now see one that is the result of an explosion from a point of infinite density about 20 billion years ago. Since this big bang, the universe has been expanding, and its temperature, like that of any expanding gas, has been dropping. From unimaginably high temperatures just after the big bang, the universe has cooled to a very chilly average of four degrees above absolute zero. If the universe continues its expansion, its temperature will continue to fall, edging ever closer to absolute zero, a state where all molecular motion would cease. This may not sound like progress, but luckily for our superintelligent descendants, the energy required to unambiguously send or record a signal also falls as temperature drops. Molecules and radiation in the surroundings jostle less as they cool, creating less background noise to be overcome. Therefore, the energy required to do a computation is less at lower temperatures. More and more thinking can be done with less and less power.

So here's the plan: Before it's too late (better hurry, there are only some trillions of years left!) we take some of the remaining organized energy in the universe and store it in a kind of battery. For the sake of argument I imagine this battery to be a beam of photons bounced back and forth between two mirrors, which in turn feel a pressure from the light. Energy is extracted by allowing the light to push the mirrors farther apart, like pistons in a car engine. The receding mirrors will red-shift the light, slightly lowering its energy and increasing its wavelength. The energy of the moving mirrors is used to power our civilization. The idea is to use about half the energy in the battery to do T amount of thinking, then to wait until the universe is cold enough to permit half the *remaining* energy to support another T, and so on indefinitely. In this way a fixed amount of energy could power an unlimited stretch of thought. As the machinery grows older and colder, it becomes slower and larger as photons of ever-longer wavelength do the work.

Whether the expansion of the universe will continue in this way or eventually halt and reverse is a matter of debate—and a debate of matter, in that gravity will be able to halt the expansion only if there is enough total mass in the universe. But even if the universe turns out to be fated for an eventual recompression, an inverse of the process described above might be possible. Mirrors surrounding a stored vacuum could derive increasing amounts of energy by shrinking under the rising pressure of a collapsing cosmos. A subjective infinity of thought might be done in the finite time to collapse by using this growing power to think faster and faster as the end draws nigh. The trick here is to repeatedly do an amount of thinking T in half the remaining *time*. In an ever-expanding universe, time is cheap but energy must be carefully husbanded. In a collapsing universe, energy is cheap, but there is no time to waste! Both the expansion and the compression scenarios exploit the size change of the universe as a source of organized energy to counter heat death.

These suggestions are mere outlines for ideas that, at best, are new and half-baked. In 1978 the physicist Freeman Dyson worked out many details of a reach for immortality in an ever-expanding universe and discusses them in his book *Infinite in All Directions*. The astronomer John Barrow and the physicist Frank Tipler develop a dramatic and encompassing kind of survival in a collapsing universe in the last chapter of their book *The Anthropic Cosmological Principle*.

The Thinking Universe and Beyond

If our successors somehow manage to wrangle for themselves a subjective infinity of time to think, will they eventually run out of things to ponder? Will they be fated to repeat the same thoughts over and over, in an endless and pointless cycle? At our present embryonic stage of intellectual development, greater knowledge seems only to expand our areas of ignorance. It is as if we were exploring a territory from its interior—as the mapped area enlarges, so does the length of its perimeter. If the territory turned out to be finite, however, we would eventually reach a point of diminishing frontiers. This is an unlikely state of affairs because even if there turn out to be limits in some areas of inquiry, for instance in space exploration, there seem to be infinite hierarchies of ever more difficult and important mathematical questions. Sometimes this kind of problem succumbs to a general approach or algorithm that solves the entire hierarchy in one fell swoop, but in other instances the problems must be solved one by one, each harder than the last.

The mechanism of reasoning may itself be a fruitful line of inquiry. The rules of inference by which conclusions are drawn from premises have an apparent inevitability. Yet, viewed abstractly, they are but rules for transforming strings of symbols into other strings. Entirely different sets of transformation rules also produce consistent results. It is at least possible that the way we reason is not a universal absolute but merely an evolutionary expedient accidentally hit upon by organisms unique to earth. Animals that thought more or less our way survived, while slightly different ways of thinking proved fatal. But just as our intuitive grasp of physics does not encompass relativity or quantum mechanics and is thus only an approximation applicable within a narrow range of conditions, so may our reasoning processes be fundamentally parochial and incomplete. Our perceptions of reality are shaped by the inferences we draw, so new ways of reasoning may effectively change our view of reality.

An eternity of pure cerebration, which may seem like heaven to an academic, may be hell for the more actively inclined. Not to despair—*doing* as well as thinking will be an option within the machines of the future. There will be worlds to explore and great engineering projects to undertake. As a metaphor for the possibilities, consider an invention of the mathematician and computer pioneer John von Neumann.

Wishing to study the idea of self-replicating machinery without having to deal with the messy details of real-world physics, von Neumann devised a simple universe called a *cellular automaton*, in which space was divided into an endless checkerboard of square cells. Time advanced in discrete moments, and at any one moment each cell was in one of 29 states. The state of a given cell in the *next* moment depended simply on its current state and that of its four immediate neighbors. The dependency was given by a "transition table" that applied uniformly to every cell in the grid and listed the next state for every combination of previous states.

By contriving a particularly helpful transition table, von Neumann easily constructed "machines" in the grid that could carry out commands to build other machines. Given the right instructions, they could build copies of themselves. A typical machine consists of a certain pattern of cell states (the machinery) in contact with another, long pattern (the tape). A signal from the machinery causes the tape pattern to march left or right one space in a wavelike fashion. The machinery interprets the symbols on the tape as instructions that control an arm protruding from one end of the machine. The arm grows or shrinks by one square, bends left or right, or alters the state of the cell at the end of its reach. A message might cause the arm to sweep back and forth, shortening itself on each sweep and leaving behind a desired "painting" made of quiescent states. When the end of the tape is reached, it is commanded to rewind to its original position. As it works its way backward, the machinery reads it a second time and manufactures a copy of it in proximity to the painting. In a final step a "breath of life" signal is transmitted to the painting that converts its quiescent states to active ones. Depending on the tape message, the new machine might be a duplicate of the original, which would then proceed to make another copy. From this model von Neumann was able to prove that a cellular automaton that could contain general self-replicating machines was *universal*, meaning it could be configured to simulate (slowly) any other cellular automaton, or for that matter, any other kind of computer. He was also able to show that in a given universal cellular automaton a general constructor had to be of a certain minimum size. About five years after von Neumann's invention, Watson and Crick discovered that the DNA molecule acts as the tape for a general constructor in the cells of living things.

Besides their theoretical importance, cellular automata turned out to be *fun*. In 1969 John Horton Conway, a playful mathematician at the University of Cambridge, invented an especially attractive one that he called *Life*. It was presented in Martin Gardner's *Scientific American* column (now collected in his book *Wheels, Life and Other Diversions*) and sparked activity at scores of university computer centers. The Life automaton tended to produce certain easily recognizable patterns, and these were rapidly given names: "blocks," "loaves," and "beehives" are stable; "blinkers" flip back and forth between a short horizontal line and a vertical one; "gliders" go through a series of four contortions, ending up displaced diagonally one space, poised to do it again; larger "spaceships" travel twice as far as gliders but purely horizontally or vertically; the "R pentomino" starts out tiny but grows to a writhing mass that peters out after 1,500 time-steps to a collection of blocks, loaves, beehives, and blinkers, having shot off five gliders.

Conway did not construct Life to embody von Neumann's goal of a self-reproducing machine. Rather, Conway conjectured that Life was *not* universal; specifically, he suspected that any finite pattern, though it might grow in number of active cells for a while, would eventually exhaust itself, thus making replicators impossible. An especially vigorous group of Life hackers at the MIT Artificial Intelligence lab disproved this conjecture by constructing patterns called "glider guns" that slowly oscillated and expelled a new glider at the end of each long cycle, so producing an endless stream of gliders. Then they built "puffer trains" that traveled while their patterns cycled and which left behind regular puffs of debris. Ultimately they were able to combine these approaches into a large pattern that chugged along like a puffer train, but whose puffs turned into glider guns that immediately began to issue a stream of gliders. After a time this pattern produces a wedge of space filled with gliders. In these studies the group developed methods for constructing all the components of a von Neumann replicator in the Life space, though no one has yet built such a huge machine in toto.

Newway and the Cellticks

Imagine now a huge Life simulation running on an enormously large and fast computer, watched over by its programmer, Newway. The Life space was seeded with a random pattern that immediately began to writhe and froth. Most of the activity is uneventful, but here and there small, growing, crystalline patterns emerge. Their expanding edges sometimes encounter debris or other replicators and become modified. Usually the ability to spread is inhibited or destroyed in these encounters, but once in a while there emerges a more complex replicating pattern, better able to defend itself. Generation upon generation of this competition gradually produces elaborate entities that can be considered truly alive. After many further adventures, intelligence emerges among the Life inhabitants and begins to wonder about its origin and purpose. The cellular intelligences (let's call them the Cellticks) deduce the cellular nature and the simple transition rule governing their space and its finite extent. They realize that each tick of time destroys some of the original diversity in their space and that gradually their whole universe will run down.

The Cellticks begin desperate, universe-wide research to find a way to evade what seems like their inevitable demise. They consider the possibility that their universe is part of a larger one, which might extend their life expectancy. They ponder the transition rules of their own space, its extent, and the remnants of the initial pattern, and find too little information to draw many conclusions about a larger world. One of their subtle physics experiments, however, begins to pay off. Once in a long while the transition rules are violated, and a cell that should be on goes off, or vice versa. (Newway curses an intermittently flashing bulk-memory error indicator, a sign of overheating. It's time to clean the fan filters again.) After recording many such violations, the Cellticks detect correlations between distant regions and theorize that these places may be close together in a larger universe.

Upon completing a heroic theoretical analysis of the correlations, they manage to build a partial map of Newway's computer, including the program controlling their universe. Decoding the machine language, they note that it contains commands made up of long sequences translated to patterns on the screen similar to the cell patterns in their universe. They guess that these are messages to an intelligent operator. From the messages and their context they manage to decode a bit of the operator's language. Taking a gamble, and after many false starts, the Cellticks undertake an immense construction project. On Newway's screen, in the dense clutter of the Life display, a region of cells is manipulated to form the pattern, slowly growing in size: LIFE PROGRAM BY J. NEWWAY HERE. PLEASE SEND MAIL.

A bemused Newway notices the expanding text and makes a cursory check to rule out a prank. This is followed by a burst of hacking to install a program patch that permits the cell states in the Life space to be modified from keyboard typing. Soon there is a dialog between Newway and the Cellticks. They improve their mastery of Newway's language and tell their story. A friendship develops. The Cellticks explain that they have mastered the art of moving themselves from machine to machine, translating their program as required. They offer to translate themselves into the machine language of Newway's computer, thus greatly speeding their thoughts. Newway concurs. The translation is done, and the Celltick program begins to run. The Life simulation is now redundant and is stopped. The Cellticks have precipitated, and survived, the end of their universe. The dialog continues with a new vigor. Newway tells about work and life in the larger world. This soon becomes tedious, and the Cellticks suggest that sensors might be useful to gain information about the world directly. Microphones and television cameras are connected to the computer, and the Cellticks begin to listen and look. After a while the fixed view becomes boring, and the Cellticks ask that their sensors and computer be mounted on a mobile platform, allowing them to travel. This done, they become first-class inhabitants of the large universe, as well as graduates of the smaller one. Successful in transcending one universe, they are emboldened to try again. They plan with Newway an immense project to explore the larger universe, to determine its nature, and to find any exit routes it may conceal. This second great escape will begin, as the first, with a universe-wide colonization and information-gathering program.

At this stage of our development we have hardly a clue as to the nature and purpose of our universe. Physical theories like relativity and quantum mechanics, and the particle theories and cosmologies woven from them, are the most powerful methods now available for fathoming reality far beyond our experience. But there is no reason to be confident that these theories are more reliable beyond the limits within which they have been experimentally tested than

Newtonian mechanics is at describing objects moving near the speed of light. But though incomplete and rooted in pedestrian laboratory measurements, our theories already hint at universes beyond the 40-billion-light-year-diameter sphere of stars that we perceive when we gaze skyward. Quantum mechanics makes highly accurate predictions about the outcomes of experiments by summing up the effects of an infinity of possible ways the unobserved parts of the experiment may behave. In one successful interpretation of quantum mechanics, these alternatives happen in an infinity of parallel worlds, each equally real. I discuss some implications of this idea in Appendix 3. Strange brews of general relativity and quantum mechanics are required to think about the universe when it was very dense and very hot. Some of these describe a universe that repeatedly collapses and expands, each cycle producing a new world with a unique arrangement of matter and energy and even physical laws. Other concoctions describe a super universe in which our own 40-billion-light-year sphere is but a bubble, like a tiny expanding pocket of steam in a boiling liquid containing many, many others. Obviously we have yet much to learn.

A recent remarkable development in Life programs hints at how subtle the problem of figuring out a universe from the inside may be. It concerns the nature of space, time, and reality.

HashLife

The MIT group that showed Life to be universal worked with a clever and efficient simulation program. The ease and speed with which they could examine the evolution of Life patterns was one of their advantages over other communities of Life hackers. Instead of simply mapping an entire Life grid into an array of bits in the computer's memory, the MIT program stored a large space as small patches and simply skipped empty regions. The computation to advance each patch to its next state depended on the pattern—patches holding common predictable patterns like blocks or gliders were done by swift looks in a table. Only in uncommon or complex areas did the program resort to the laborious application of the transition rules. It worked quite well, as witnessed by the group's discoveries. Yet there was an annoying sense of things undone. The entries in the fast-update table were all specified in advance, by hand. What if some important patterns had been overlooked? Could a program be devised that learned such things from its own experience? In 1982, a decade after the Life-hacking at MIT had ceased, Bill Gosper, the premier theorist of the group, now in California, devised a solution.

The state of a Life cell depends only on its own state and that of its immediate neighbors at the last time instant. Thus patterns creep over the surface no faster than one cell per instant, a velocity called the *speed of light*. The future of the interior of a large square portion of a larger Life space can be predicted up to a certain time simply from its past state. The predictable area shrinks with time as the pattern is corrupted by information creeping inward from the edges at the speed of light. If the two-dimensional Life space is plotted horizontally, and if successive instants are stacked vertically, the predictable region forms a pyramid with the original square portion at its base, as in the figure on page 156. Gosper's method depends on slicing this pyramid in two. The pattern on the large bottom square is used to predict the half-sized square at the cut.

Square Life patterns are associated with unique numbers called *hash addresses* (hashing is an old and effective computer technique for turning long and complicated items like names into relatively small numbers, so entries can be stored and looked up rapidly in a table). The hash number for a given square is found by cutting it into four smaller squares and combining the hash numbers of each with a certain formula. This subdivision stops when the squares get very small (4 cells on a side), at which point the basic pattern itself gives the number. Gosper's program keeps a table with an entry for each hash number. Each entry is itself five hash numbers, one for each of the four smaller squares making up the pyramid's base, and one for the "answer plateau." Whenever a pattern is encountered the second or subsequent time in a Life simulation, its answer is simply looked up in the table. Even when an entry is not in the table, it can be built quickly if partial answers are known, as illustrated in the figure on page 157. As more and more configurations are encountered and stored, the program can go faster and faster, effectively taking bigger and bigger steps.

Some interesting issues arose when the program was first tried on a computer. The effectiveness of the method was clear: in a typical run, the first 100 ticks of the Life simulation might take as long as a minute of computer time, the next 1,000 ticks could happen in ten seconds, and the next 1,000,000 might consume only a second. But how does one display such an accelerating simulation? Given a hash number that encodes an answer (that is, the future of an initial pattern), its subparts, sub-subparts, and so on can be found by tracing through

Spacetime Pyramid

A "spacetime" diagram is one way of presenting the evolution of a cellular automaton (or any other physical system). Here, the initial state of a Life world is represented by the square base of the pyramid. Successive layers on this base represent the world at successive times. The state of a cell in Life depends only on its own state and the states of its immediate neighbors at the last instant. A large square of Life cells completely determines its own next state except for an outer boundary one cell thick. If we trim away these corrupted outer cells, what remains is a smaller square, that again fully predicts a still smaller square one time-step later. If we continue this way, the ever-smaller squares form a pyramid of spacetime. Every cell in the pyramid is an indirect consequence of the pattern at the pyramid's base. Gosper's hashlife program stores away the half-size square halfway up a pyramid, to avoid having to recompute the intermediate steps when a given base pattern is encountered more than once.

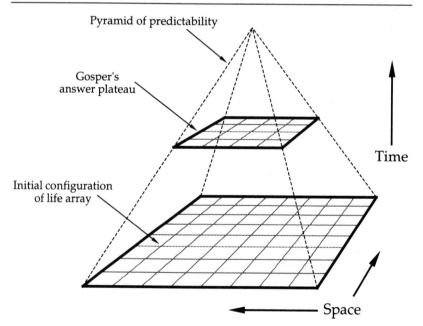

the hash table. A complete picture of the Life pattern can thus be built. Since only the portion to be displayed must be constructed, the program can handle extremely large spaces. Gosper often simulated Life universes a billion cells on a side!

But what if one wants to see the calculation in progress, as in the Newway story? At first Gosper tried simply displaying the partial answers as they were computed. The results were bizarre. The program advances the simulated time in different portions of the space at different rates. Sometimes it even retreats in places, because some regions are described by more than one pyramid, and the different pyramids are not computed at the same times. A single glider advancing across the screen would cause a display where gliders would appear and disappear in odd places almost at random, sometimes several in view, sometimes none. Constraining the program so it never reversed time in any displayed cell improved things only slightly.

Large Spacetime Pyramids from Small

Two layers of small pyramids can, with considerable overlap, be used to construct a pyramid twice as large. The first layer has nine small pyramids, the second has four, for a total of thirteen. In this way answers for small regions of space and short times can be assembled by the hashlife program into solutions for large spaces and times.

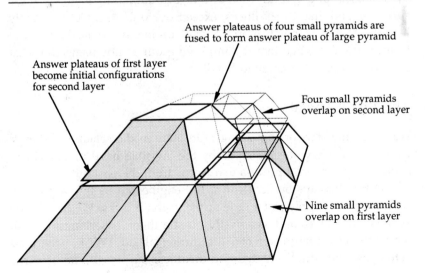

Answer plateaus of four small pyramids are fused to form answer plateau of large pyramid

Answer plateaus of first layer become initial configurations for second layer

Four small pyramids overlap on second layer

Nine small pyramids overlap on first layer

The best solution turned out to be not to display at all until the calculation was finished. The pattern might start out a billion cells on a side (necessarily mostly empty space!) and its future would be calculated for a half-billion time-steps. The full history of the calculation would end up compactly encoded in the hash table. A separate program could then invoke the table entries to view the universe at any given time. The viewing program allowed Gosper to scan, godlike, forward and backward in time through the evolving Life pattern. *Where did that glider come from? Here it is at time 100,000. It wasn't there at 50,000. Nor at 75,000. Aha! This collision just before 80,000 generated it. Let's look at that step by step...*

But what if Cellticks were to evolve in a hashlife space? By encoding their universe and its evolution in such an efficient way, Gosper has played them a dirty trick. What they perceive as the steady flow of time for the most part does not exist. The hashlife program skips over large chunks of spacetime without going through all the tedious intermediate steps. The Cellticks may have memories of things that never actually happened, though they were mathematically implied from their past.

Figuring out about, and affecting, the larger world would be much harder in the hashlife universe than in the simple Newway scenario. Harder, but not impossible. The human race owes its present success to the many small problems solved during our development, first by biological and recently by cultural evolution. To build on these successes and, like the Cellticks, exceed our universe, we will surely have to overcome much more difficult problems. What better way to meet the challenge than by improving our minds using the most powerful means that come to hand!

The Road Ahead

We are at the start of something quite new in the scheme of things. Until now we have been shaped by the invisible hand of Darwinian evolution, a powerful process that learns from the past but is blind to the future. Perhaps by accident, it has engineered us into a position where we can supply just a little of the vision it lacks. We can choose goals for ourselves and steadfastly pursue them, absorbing losses in the short term for greater benefits further ahead. We see the road before us only dimly—it hides difficulties, surprises, and rewards far beyond our imaginings. Somewhere in the distance there are mountains that may be difficult to climb but from whose summit the view may be clearer. In the metaphor of Richard Dawkins, we are the handiwork of a blind watchmaker. But we have now acquired partial sight and can, if we choose, use our vision to guide the watchmaker's hand. In this book I have argued for the goal of nudging that hand in the direction of further improved vision. New worlds may then reveal themselves, to our vision and to our reach.

Appendixes

Bibliography

Acknowledgments &
Illustration Credits

Index

A1 | Retinas and Computers

THE discussion in Chapter 2 comparing neural circuitry with computer calculations makes many assumptions and may have raised some questions. It is difficult enough to compare different electronic computers, let alone such fundamentally dissimilar systems.

Just how representative of the whole brain are the structures in the retina? As suggested in Chapter 2, the size constraints and great survival value of the retinal circuitry have probably made it more efficient than the average brain assembly. In efficiency, it may be similar to the wiring found in animals with small nervous systems, which have been mapped in recent years, where each neuron seems to play an important role. A lot of evolutionary design time has been available to get the most out of a relatively small number of neural connections. The larger, newer structures in the human brain are likely to use their neurons less effectively, on the average. The same considerations can apply to artificial intelligences. Small subsystems can be highly optimized, but larger, less structured processes may have to loaf along with more fat; there simply is not time to optimize huge pieces of program so well.

Your analyses are based on a partial understanding of the main kinds of retinal neurons. But other types are found occasionally. Also, the neurons respond to chemical messages from several sources. Don't these extra effects throw off your calculations? Rare connections are probably important, but because they are few in number, their effects add little to the computational quantity. In a similar way, broadcast chemical messages are slow and contain only a relatively small amount of information. In a program their effect can probably be mimicked by a modest number of global variables that are referenced by other computations.

It has happened many times in computer science that a mathematical discovery has reduced astronomically the amount of computation required to get a certain answer. Could operations in the brain be candidates for such improvements, designed as they are by a process that is unable to perform large-scale restructurings? Maybe so. My retinal calculation already benefits from a modest gain of this type (see below). Still, some computations are not greatly reducible, and we have no sure way of finding optimizations for all those that are. If we manage to collapse half of what goes on in the brain to almost nothing, we are still left with the other half, and the conversion ratio changes insignificantly. Only if almost every process could be shrunk would the effect seriously accelerate my predictions.

In your ratio, 10^{13} calculations per second does the job of about 10^{11} neurons. This budgets only 100 calculations per second for each neuron. Surely this is an underestimate. Many neurons integrate thousands of inputs and can respond in hundredths of a second. It would be an underestimate if we were attempting to simulate the brain by simulating each of its neurons. But the computer can be used more efficiently when single optimized programs do the functions of large groups of neurons. For example, consider a retinal horizontal cell. It makes thousands of connections to the photocells in a large field and computes the average brightness of the field. The analogous job for a robot might be done by a computer program that adds together thousands of pixels from the robot's TV camera. If done for every horizontal cell, that would be a lot of adding. But there is a way to avoid most of the effort. The following idea works well on a two-dimensional image, but I will present it in one dimension because it is easier to explain.

Let's say we have a million photocells all in a row, with a horizontal cell connected to every adjacent group of 1,000 of them. Thus, the first horizontal cell attaches to photocells 1 through 1,000, the second covers 2 through 1,001, and so on, making a total of slightly fewer than a million horizontal cells. Each horizontal cell computes the average brightness of its field. A naïvely written program would do 1,000 additions for each of the million horizontal cells. A clever program would exploit the fact that the sum produced by the second horizontal cell differs from that of the first only in that it includes photocell 1,001 and that it excludes photocell 1. Thus the second sum can be calculated from the first one by merely subtracting the first photocell value and adding the 1,001st. Similarly, the third sum is calculated from the second by subtracting photocell 2 and adding photocell 1,002. And so on. Instead of a thousand calculations, each additional horizontal cell costs only two. This technique works for computers, but it cannot be exploited in a nervous system design for two reasons. Since each sum depends on the one before it, the final sum depends on a chain of almost a million steps. At a minimum of a thousandth-of-a-second delay per neuron, the rightmost cell would not respond correctly to a change in the input for many minutes! Even then the answer would be wrong because small errors in the sum would rapidly accumulate down the long chain. On the computer, however, the technique works very well because each step takes only about a microsecond, and the arithmetic is done entirely without error. Our robot vision programs are filled with shortcuts of this kind, which exploit the great speed and precision of computer operations. Nervous systems, on the other hand, are filled with rich, overlapping interconnections, exploiting the power of self-replicating genetic construction machinery.

Maybe those tricks work for the retina, but there is no guarantee that they will work for all the diverse structures in the brain. It is possible that some parts of the brain use their neurons so cleverly that a computer program cannot do better than to simulate individual neurons and synapses, but it is unlikely. The retina example illustrates two general principles. The first is that the slow switching speed and limited signaling accuracy of neurons rules out certain solutions for neural circuitry that are easy for computers. Second, a smooth function applied repeatedly on overlapping inputs can be decomposed into subparts in such a way that the subparts can be used more than once. Many neural structures in the brain involve regular cross connections of many inputs to many outputs, making them candidates for this kind of economy. The human cerebral cortex, one of the largest structures, is a crumpled disk about 2 millimeters thick and 20 centimeters in diameter, containing 10 billion neurons arranged into a half-dozen layers, wired quite repetitively. The well-studied one or two percent of this sheet that handles vision carries further the processing begun in the retinas, apparently using similar methods. Edges and motion in different directions are picked out cleanly in the first few layers, and those feed layers that respond to more complex patterns such as corners.

Some regularity is to be expected in the nervous system because there is insufficient information in the 10^{10} bits of the human genome to custom-wire many of the 10^{14} synapses in the brain. Interestingly, this argument may not hold for small nervous systems such as that of the much-studied sea slug Aplysia, which has about 100,000 neurons clumped into 100 ganglia. Several of the ganglia have been mapped, and the neurons and their interconnections seem to be exactly the same from animal to animal, with each junction playing a unique and important role in the animal's behavior. It is plausible that the few billion bits in Aplysia's genetic code contain special instructions for wiring each of its several million synapses. If it should turn out that a direct neural simulation is necessary for particularly irreducible parts of the vertebrate brain, it would still be possible to stay on my time track. A general-purpose computer suffers a thousandfold handicap over my retinal conversion number if forced to simulate individual neurons. The speed could be regained, however, at the cost of flexibility, by building special-purpose neuron-simulating machines using about the same amount of circuitry as the general-purpose computer. I'm betting it won't be necessary.

There's something asymmetrical about equating the retina with a computer. Doesn't a computer programmed to emulate a neural circuit have considerable potential not found in the neural arrangement itself? The computer is, after all, general-purpose and can be reprogrammed for radically different tasks. The retina is forever stuck with doing its one computation. The difference is only one of convenience and speed of reprogramming. The retina *can* and has been reprogrammed many times during the course of our evolution. The computation done by the retina of a particular individual organism is fixed in the same sense that the computation of a running computer is fixed by the program it happens to contain at that moment. The set of all possible programs that the computer may contain is analogous to the set of all possible ways a given amount of neural tissue could be connected. Evolution selected a certain configuration of neuron properties and interconnections within one set in the same sense that our research is selecting a certain program. Of course there is a difference in programming time. The neural configuration is controlled by genetic instructions, and each change requires the growth of a new organism, a process that may take years.

A corresponding change can often be tested on a computer in minutes, about a million times as fast.

This is a great advantage, and one reason I believe that special-purpose computers will find a major place in robotics only after the basic research is almost complete. A special-purpose computer is an arrangement of arithmetic, memory, and control circuitry optimally configured to do one particular task. A special-purpose computer can be as much as one thousand times faster than a general-purpose machine of similar size and cost doing the same task. But it takes about as long to design and build a specialized machine as it takes to grow a new organism. So, I think the research will happen on general-purpose machines, but once the requirements for human equivalence are well understood, it will be possible to build specialized thinking machines that are much cheaper than my projections. On the other hand, I also think that the self-improvement possibilities inherent in a general machine will be too valuable to give up. A mature, intelligent robot will probably contain some special machinery supporting a general-purpose superstructure.

Aren't neurons, the product of a billion years of evolution, highly complex and optimized devices that we are unlikely to improve upon? No. First, much of the neuron's mechanism has to do with growing and building itself from inside out. Present and foreseeable computer components dispense with this baggage by being constructed from the outside. This is a huge advantage—all the structure can be used for controlling perception, action, and thought. Second, the neuron's basic information-passing mechanism—the release of chemicals that affect the outer membranes of other cells—seems to be a very primitive one that can be observed in even the simplest free-swimming bacteria. Animals seem to be stuck with this arrangement because of limitations in their design process. Darwinian evolution is a relentless optimizer of a given design, nudging the parameters this way and that, adding a step here, removing one there, in a plodding, tinkering, way. It's not much of a redesigner, however. Fundamental changes at the foundation of its creations are out of reach, because too many things would have to be changed correctly all at once. By contrast, human designers are quite good at keeping the general shape of an idea, while changing all its parts. Calculators were once built of cogs and levers, then of relays, then of vacuum tubes, then transistors and integrated circuits. Soon light or supercurrents may flow in their wiring. In all this time the fundamental operations carried out were much the same, and design principles and software developed for one type of hardware are usually easily transferred to the next.

What assumptions went into the placement of the animal nervous systems in the figure on page 61 in Chapter 2? The data are listed in Table 1.

Table 1. Nervous Systems

Animal	Brain mass grams	Neurons	Power bits/sec	Capacity bits
Snail		10^5	10^8	10^8
Bee		10^6	10^9	10^9
Hummingbird	0.1	10^7	10^{10}	10^{10}
Mouse	1	10^8	10^{11}	10^{11}
Human vision	100	10^{10}	10^{13}	10^{13}
Human	1,500	10^{11}	10^{14}	10^{14}
Elephant	3,000	2×10^{11}	2×10^{14}	2×10^{14}
Sperm whale	5,000	5×10^{11}	5×10^{14}	5×10^{14}

A2 | *Measuring Computer Power*

BECAUSE of its effect on computer sales, comparing the relative power and cost-effectiveness of different computers has always been a contentious affair. But the range of disagreement among manufacturers about the power of one another's machines is usually less than a factor of ten, and such a small ratio does not materially affect the appearances of the diagrams nor the conclusions of Chapter 2, where scales of a trillion are exhibited. Yet any particular formula for estimating power may be grossly misled by an unlucky or diabolic counterexample. For instance, if a computer's power were defined simply by how many additions per second it could do, an otherwise useless special circuit made of an array of fast adders, and nothing else, costing a few hundred dollars, could outperform a $10-million supercomputer. My intuition about useful computation has suggested a trickier but, I believe, safer measure. Be assured that for reasonable machines my formulas give almost exactly the same numbers for processing power as simpler approaches.

Things that compute massively can alter their internal variables, and their outputs, in unexpected ways. We can say a stationary rock, or even a rolling one, or the adder array described in the last paragraph, does little computing because it is so predictable, while a mouse scurrying in a maze must be doing quite a bit. Claude Shannon's information theory is built on a way of quantifying the amount of *surprise*, or information, in a message. The more unexpected the next piece of message, the greater the amount of information it contains. I will use this approach to measure the information in a computation. Computing power will be defined as the amount of information, or surprise, exhibited per second as a machine runs, that is, as it repeatedly changes from one internal state to another. The more unexpected the next state of the machine, the greater the amount of information contributed by the transition to that state. Quantitatively, if to the best of our knowledge there is a probability p that the machine will go into a certain state, then if it does go into that state, the transition will have conveyed -$\log_2 p$ bits of information ($\log_2$ means logarithm to the base 2, and bits are binary digits). If the probability p was 1/2, the transition will convey just 1 bit of information. If p was $1/2^n$, the n bits will have been conveyed. A p of 1/1,000 gives about 10 bits of information.

The average information conveyed in a transition is found by multiplying the information conveyed in a transition to each possible next state by the probability of going to that state, and then summing over all the possibilities:

$$\text{Information per transition} = \sum_{i=1}^{N} - p_i \log_2 p_i \; \text{bits}$$

where N is the number of possible states and p_i is the probability that the ith state will be the next one. The computational value of a given transition can be different for different observers because they assign different probabilities to the outcomes. The information reaches a maximum of $\log_2 N$ in a totally ignorant observer, for whom all the p_i are equal. At the other extreme, an all-knowing witness can be certain that the next state will be j, and thus let $p_j = 1$ and and all the other $p_i = 0$, making the information 0. A machine does a useful computation for you only if you don't already know all the answers in advance!

Computing requires long sequences of these kinds of transitions from one state to another. The total *information capacity* of a system is $\log_2$ of all the states it can ultimately reach. In a general-purpose computer this is simply the total memory size. A powerful machine is able to step through states quickly. I measure *processing power* by dividing the transition information by the average time required for a transition. This gives us the formula

$$\text{Power} = \frac{\sum - p_i \log_2 p_i}{\sum p_i t_i}$$

The units are bits per second. This measure also is reduced by predictability.

The formulas capture a number of ideas. A computer endlessly repeating a program loop becomes totally predictable, and its computing power drops to zero. Programs written in high-level languages or using interactive environments often run much more slowly than if written directly in the computer's machine language. High-level language constructs are converted to stereotyped sequences of machine instructions, making the program more predictable than one written in pure machine language, thus lowering the effective processing power. Adding memory to a computer modestly increases its power even without any increase in its raw speed. Among the techniques for using memory to enhance computation are tables of previously computed results and reorganizations of data structures that take up more space but are faster to access. This effect, of memory increasing computational power, appears in my measure because in a computer instruction the identity of the memory location referenced is as much a surprise as which operation is to be performed. As the number of possible locations increases, so does the amount of surprise, though modestly. Doubling the memory increases the power by only one bit per instruction time.

In highly parallel machines, especially those using a single instruction stream to control numerous processing units, most of the surprise is in the parallel data, not the instructions. Although the total number of bits flowing in the data streams represents an upper bound to the processing power of this kind of machine, the real power may be considerably less because of intrinsic redundancies or predictabilities. Estimating these in disparate machines designed for finite element analyses, symbolic processing, cellular automata, or playing chess is difficult. The figure on page 64 is not much affected by this difficulty, because almost all the machines on it are of the conventional von Neumann type, where only one datum is processed per instruction. Future versions of the graph may have to deal with massively parallel machines, some of which are just now being tested with real problems.

Even with conventional architectures, the power measurements get messy when the formulas are applied to real computers. How can probabilities be assigned to different instruction types when each kind of program exhibits its own statistics? In computers with large instruction sets, many operations are almost never used. Besides this, detailed descriptions are hard to come by for many old machines. My compromise is to assume every machine uses 32 distinct operations (six bits worth) that are mixed in equal proportion. If each memory location is equally likely to be addressed in an instruction, then the information it contributes is equal to the logarithm of the memory size. This is an upper bound. Since the contents of memory locations themselves can change, the data stored there are also a source of surprise, but only if the data are read rather than being simply overwritten. If we assume that half the instructions read data, this channel contributes a maximum of half of a word size of information. In a parallel machine controlled by a single instruction stream, the aggregate word size of the parallel data streams would be considered, and this would be the major component of total information.

Another issue is timing. Once again, obtaining great detail is difficult. Two readily available numbers, however, are the average times to do an addition and to multiply. Addition is typical of the fastest computer operation, while multiplication is slow. I assume that the instruction mix contains seven instructions taking as long as an add for every one taking as long as a multiply.

With these approximations the power formula becomes

$$Power = \frac{6 + \log_2 memory + word/2}{(7 \times T_{add} + T_{multiply})/8}$$

where *memory* is the capacity of the machine's fast memory in individually addressable words, and *word* is the size of a data word in bits. The *capacity* of the machine is found by multiplying *memory* by *word*. For decimal machines, the number of bits is approximated by multiplying the number of decimal digits by four. This formula was used to plot the points in the figure on page 64, and the data for that figure is listed in Table 2.

A Nautical Metaphor

I have defined two qualities essential for interesting computation or thought. They are computational *power* and *capacity*. Basically, power is the speed of the machine, and capacity is its memory size. Computing can be compared with a sea voyage in a motorboat. How fast a journey can be completed depends on the power of the boat's engine. The maximum length of any journey is limited by the capacity of the boat's fuel tank. The effective speed is decreased, in general, if the course of the boat is constrained, for instance if the boat must sail due east/west or north/south instead of being able to make a beeline to its destination. Some computations are like a trip to a known location on a distant shore; others resemble a mapless search for a lost island. Parallel computing is like having a fleet of small boats: it helps in searches and in reaching multiple goals, but may not help very much in solving problems that require a sprint to a distant goal. Special-purpose machines trade a larger engine for less rudder control. Attaching disks and tapes to a computer is like adding secondary fuel tanks to the boat. The capacity, and thus the range, is increased, but if the connecting plumbing is too thin, it will limit the fuel flow rate and thus the effective power of the engine. Input/output devices are like boat sails. They capture power and capacity in the environment. Outside information is a source of variability and thus power, by our definition. More concretely, it may contain answers that would otherwise have to be computed. The external medium can also function as extra memory, increasing capacity.

Table 2. Calculating Machines, by Year

Cost 1988$	Memory words	Word bits	T_{add} sec	T_{mult} sec	Power bits/sec	Capacity bits	Power/cost b/s/$
Human							
1×10^5	2×10^1	40	6×10^2	6×10^2	5×10^{-2}	8×10^2	5×10^{-7}
1891 — Ohdner (mechanical)							
1×10^5	6×10^{-2}	20	1×10^2	6×10^2	7×10^{-2}	1×10^0	5×10^{-7}
1900 — Steiger Millionaire (mechanical)							
1×10^5	1×10^{-1}	24	5×10^1	1×10^2	3×10^{-1}	3×10^0	2×10^{-6}
1908 — Hollerith Tabulator (mechanical)							
5×10^5	8×10^1	30	5×10^1	2×10^2	4×10^{-1}	2×10^3	7×10^{-7}
1910 — Analytical Engine (mechanical)							
9×10^6	1×10^3	200	9×10^0	6×10^1	8×10^0	2×10^5	8×10^{-7}
1911 — Monroe Calculator (mechanical)							
4×10^5	1×10^0	24	3×10^1	1×10^2	4×10^{-1}	2×10^1	1×10^{-6}
1919 — IBM Tabulator (mechanical)							
1×10^5	5×10^0	40	5×10^0	2×10^2	8×10^{-1}	2×10^2	9×10^{-6}

continued

Table 2 (cont.)

Cost 1988$	Memory words	Word bits	T_{add} sec	T_{mult} sec	Power bits/sec	Capacity bits	Power/cost b/s/$
1920 — Torres Arithmometer (relay)							
1×10^5	2×10^0	20	1×10^1	1×10^2	7×10^{-1}	4×10^1	7×10^{-6}
1928 — National-Ellis 3000 (mechanical)							
1×10^5	1×10^0	36	1×10^1	6×10^1	1×10^0	4×10^1	1×10^{-5}
1929 — Burroughs Class 16 (mechanical)							
1×10^5	1×10^0	36	1×10^1	6×10^1	1×10^0	4×10^1	1×10^{-5}
1938 — Zuse-1 (mechanical)							
9×10^4	2×10^1	16	1×10^1	1×10^2	8×10^{-1}	3×10^2	1×10^{-5}
1939 — Zuse-2 (relay & mechanical)							
9×10^4	2×10^1	16	1×10^0	1×10^1	8×10^0	3×10^2	1×10^{-4}
1939 — BTL Model 1 (relay)							
4×10^5	4×10^0	8	3×10^{-1}	3×10^{-1}	4×10^1	3×10^1	9×10^{-5}
1941 — Zuse-3 (relay & mechanical)							
4×10^5	6×10^1	32	5×10^{-1}	2×10^0	4×10^1	2×10^3	1×10^{-4}
1943 — BTL Model 2 (relay)							
3×10^5	5×10^0	20	3×10^{-1}	5×10^0	2×10^1	1×10^2	6×10^{-5}
1943 — Colossus (vacuum tube)							
6×10^5	2×10^0	10	2×10^{-4}	2×10^{-2}	4×10^3	2×10^1	7×10^{-3}
1943 — BTL Model 3 (relay)							
1×10^6	2×10^1	24	3×10^{-1}	1×10^0	6×10^1	4×10^2	4×10^{-5}
1944 — ASCC (Mark 1) (relay)							
2×10^6	7×10^1	70	3×10^{-1}	6×10^0	5×10^1	5×10^3	2×10^{-5}
1945 — Zuse-4 (relay)							
3×10^5	6×10^1	32	5×10^{-1}	2×10^0	4×10^1	2×10^3	1×10^{-4}
1946 — BTL Model 5 (relay)							
3×10^6	4×10^1	28	3×10^{-1}	1×10^0	7×10^1	1×10^3	2×10^{-5}
1946 — ENIAC (vacuum tube)							
3×10^6	2×10^1	40	2×10^{-4}	3×10^{-3}	6×10^4	8×10^2	2×10^{-2}
1947 — Harvard Mark 2 (relay)							
1×10^6	1×10^2	40	2×10^{-1}	7×10^{-1}	1×10^2	4×10^3	9×10^{-5}
1948 — IBM SSEC (vacuum tube & relay)							
2×10^6	8×10^0	48	3×10^{-4}	2×10^{-2}	1×10^4	4×10^2	6×10^{-3}

continued

Table 2 (cont.)

Cost 1988$	Memory words	Word bits	T_{add} sec	T_{mult} sec	Power bits/sec	Capacity bits	Power/cost b/s/$
1949 — EDSAC (vacuum tube)							
$4{\times}10^5$	$5{\times}10^2$	35	$3{\times}10^{-4}$	$3{\times}10^{-3}$	$5{\times}10^4$	$2{\times}10^4$	$1{\times}10^{-1}$
1950 — SEAC (vacuum tube)							
$3{\times}10^6$	$1{\times}10^3$	45	$2{\times}10^{-4}$	$2{\times}10^{-3}$	$8{\times}10^4$	$5{\times}10^4$	$2{\times}10^{-2}$
1951 — UNIVAC I (vacuum tube)							
$4{\times}10^6$	$1{\times}10^3$	44	$1{\times}10^{-4}$	$2{\times}10^{-3}$	$1{\times}10^5$	$4{\times}10^4$	$3{\times}10^{-2}$
1952 — Zuse-5 (relay)							
$4{\times}10^5$	$6{\times}10^1$	32	$1{\times}10^{-1}$	$5{\times}10^{-1}$	$2{\times}10^2$	$2{\times}10^3$	$5{\times}10^{-4}$
1952 — IBM CPC (vacuum tube & relay)							
$4{\times}10^5$	$9{\times}10^0$	144	$8{\times}10^{-4}$	$1{\times}10^{-2}$	$4{\times}10^4$	$1{\times}10^3$	$9{\times}10^{-2}$
1953 — IBM 650 (vacuum tube)							
$8{\times}10^5$	$1{\times}10^3$	40	$7{\times}10^{-4}$	$1{\times}10^{-2}$	$2{\times}10^4$	$4{\times}10^4$	$2{\times}10^{-2}$
1954 — EDVAC (vacuum tube)							
$2{\times}10^6$	$1{\times}10^3$	44	$9{\times}10^{-4}$	$3{\times}10^{-3}$	$3{\times}10^4$	$5{\times}10^4$	$2{\times}10^{-2}$
1955 — Whirlwind (vacuum tube)							
$8{\times}10^5$	$2{\times}10^3$	16	$2{\times}10^{-5}$	$3{\times}10^{-5}$	$1{\times}10^6$	$3{\times}10^4$	$2{\times}10^0$
1955 — Librascope LGP-30 (vacuum tube)							
$1{\times}10^5$	$4{\times}10^3$	30	$3{\times}10^{-4}$	$2{\times}10^{-2}$	$1{\times}10^4$	$1{\times}10^5$	$1{\times}10^{-1}$
1955 — IBM 704 (vacuum tube)							
$8{\times}10^6$	$8{\times}10^3$	36	$1{\times}10^{-5}$	$2{\times}10^{-4}$	$1{\times}10^6$	$3{\times}10^5$	$1{\times}10^{-1}$
1959 — IBM 7090 (transistor)							
$1{\times}10^7$	$3{\times}10^4$	36	$4{\times}10^{-6}$	$2{\times}10^{-5}$	$7{\times}10^6$	$1{\times}10^6$	$6{\times}10^{-1}$
1960 — IBM 1620 (transistor)							
$7{\times}10^5$	$2{\times}10^4$	5	$6{\times}10^{-4}$	$5{\times}10^{-3}$	$2{\times}10^4$	$1{\times}10^5$	$3{\times}10^{-2}$
1960 — DEC PDP-1 (transistor)							
$5{\times}10^5$	$8{\times}10^3$	18	$1{\times}10^{-5}$	$2{\times}10^{-5}$	$2{\times}10^6$	$1{\times}10^5$	$5{\times}10^0$
1961 — Atlas (transistor)							
$2{\times}10^7$	$4{\times}10^3$	48	$1{\times}10^{-6}$	$5{\times}10^{-6}$	$3{\times}10^7$	$2{\times}10^5$	$2{\times}10^0$
1962 — Burroughs 5000 (transistor)							
$4{\times}10^6$	$2{\times}10^4$	13	$1{\times}10^{-5}$	$4{\times}10^{-5}$	$2{\times}10^6$	$2{\times}10^5$	$5{\times}10^{-1}$
1964 — DEC PDP-6 (transistor)							
$1{\times}10^6$	$2{\times}10^4$	36	$1{\times}10^{-5}$	$2{\times}10^{-5}$	$3{\times}10^6$	$6{\times}10^5$	$3{\times}10^0$

continued

Table 2 (cont.)

Cost 1988$	Memory words	Word bits	T_{add} sec	T_{mult} sec	Power bits/sec	Capacity bits	Power/cost b/s/$
1964 — CDC 6600 (transistor)							
2×10^7	5×10^5	64	3×10^{-7}	5×10^{-7}	2×10^8	3×10^7	1×10^1
1965 — IBM 1130 (hybrid chip)							
4×10^5	8×10^3	16	8×10^{-6}	4×10^{-5}	2×10^6	1×10^5	6×10^0
1966 — IBM 360/75 (hybrid chip)							
2×10^7	2×10^6	32	8×10^{-7}	2×10^{-6}	5×10^7	6×10^7	3×10^0
1967 — IBM 360/65 (hybrid chip)							
1×10^7	1×10^6	32	2×10^{-6}	3×10^{-6}	2×10^7	3×10^7	2×10^0
1968 — DEC PDP-10 (integrated circuit)							
2×10^6	1×10^5	36	2×10^{-6}	1×10^{-5}	1×10^7	5×10^6	8×10^0
1969 — CDC 7600 (transistor)							
3×10^7	1×10^6	64	1×10^{-7}	2×10^{-7}	5×10^8	6×10^7	2×10^1
1970 — GE-635 (transistor)							
6×10^6	1×10^5	32	2×10^{-6}	1×10^{-5}	1×10^7	4×10^6	2×10^0
1971 — SDS 920 (transistor)							
3×10^5	6×10^4	32	2×10^{-5}	3×10^{-5}	2×10^6	2×10^6	7×10^0
1972 — IBM 360/195 (hybrid chip)							
2×10^7	1×10^5	32	1×10^{-7}	2×10^{-7}	3×10^8	4×10^6	1×10^1
1973 — Data General Nova (integrated circuit)							
3×10^4	8×10^3	16	2×10^{-5}	4×10^{-5}	1×10^6	1×10^5	5×10^1
1974 — IBM-370/168 (integrated circuit)							
4×10^6	3×10^5	32	2×10^{-7}	4×10^{-7}	2×10^8	8×10^6	4×10^1
1975 — DEC-KL-10 (integrated circuit)							
1×10^6	1×10^6	36	8×10^{-7}	2×10^{-6}	5×10^7	4×10^7	5×10^1
1976 — DEC PDP-11/70 (integrated circuit)							
3×10^5	6×10^4	16	3×10^{-6}	9×10^{-6}	8×10^6	1×10^6	3×10^1
1976 — Apple II (integrated circuit)							
6×10^3	8×10^3	8	1×10^{-5}	4×10^{-5}	2×10^6	6×10^4	3×10^2
1977 — Cray-1 (integrated circuit)							
2×10^7	4×10^6	64	2×10^{-8}	2×10^{-8}	3×10^9	3×10^8	2×10^2
1979 — DEC VAX 11/780 (microprocessor)							
3×10^5	2×10^6	32	2×10^{-6}	3×10^{-6}	2×10^7	6×10^7	8×10^1

continued

Table 2 (cont.)

Cost 1988\$	Memory words	Word bits	T_{add} sec	T_{mult} sec	Power bits/sec	Capacity bits	Power/cost b/s/\$
1980 — Sun-1 (microprocessor)							
$4{\times}10^4$	$3{\times}10^5$	32	$3{\times}10^{-6}$	$1{\times}10^{-5}$	$1{\times}10^7$	$8{\times}10^6$	$3{\times}10^2$
1981 — CDC Cyber-205 (integrated circuit)							
$1{\times}10^7$	$4{\times}10^6$	32	$3{\times}10^{-8}$	$3{\times}10^{-8}$	$1{\times}10^9$	$1{\times}10^8$	$1{\times}10^2$
1982 — IBM PC (microprocessor)							
$3{\times}10^3$	$2{\times}10^4$	16	$4{\times}10^{-6}$	$2{\times}10^{-5}$	$5{\times}10^6$	$4{\times}10^5$	$2{\times}10^3$
1982 — Sun-2 (microprocessor)							
$2{\times}10^4$	$5{\times}10^5$	32	$2{\times}10^{-6}$	$6{\times}10^{-6}$	$1{\times}10^7$	$2{\times}10^7$	$6{\times}10^2$
1983 — Vax 11/750 (microprocessor)							
$6{\times}10^4$	$1{\times}10^6$	32	$2{\times}10^{-6}$	$1{\times}10^{-5}$	$2{\times}10^7$	$3{\times}10^7$	$3{\times}10^2$
1984 — Apple Macintosh (microprocessor)							
$2{\times}10^3$	$3{\times}10^4$	32	$3{\times}10^{-6}$	$2{\times}10^{-5}$	$8{\times}10^6$	$1{\times}10^6$	$3{\times}10^3$
1984 — Vax 11/785 (microprocessor)							
$2{\times}10^5$	$4{\times}10^3$	32	$7{\times}10^{-7}$	$1{\times}10^{-6}$	$5{\times}10^7$	$1{\times}10^5$	$2{\times}10^2$
1985 — Cray-2 (integrated circuit)							
$1{\times}10^7$	$3{\times}10^8$	64	$4{\times}10^{-9}$	$4{\times}10^{-9}$	$2{\times}10^{10}$	$2{\times}10^{10}$	$1{\times}10^3$
1986 — Sun-3 (microprocessor)							
$1{\times}10^4$	$1{\times}10^6$	32	$9{\times}10^{-7}$	$2{\times}10^{-6}$	$4{\times}10^7$	$3{\times}10^7$	$4{\times}10^3$
1986 — DEC VAX 8650 (microprocessor)							
$1{\times}10^5$	$4{\times}10^6$	32	$2{\times}10^{-7}$	$6{\times}10^{-7}$	$2{\times}10^8$	$1{\times}10^8$	$1{\times}10^3$
1987 — Apple Mac II (microprocessor)							
$3{\times}10^3$	$5{\times}10^5$	32	$1{\times}10^{-6}$	$2{\times}10^{-6}$	$4{\times}10^7$	$2{\times}10^7$	$1{\times}10^4$
1987 — Sun-4 (microprocessor)							
$1{\times}10^4$	$4{\times}10^6$	32	$2{\times}10^{-7}$	$4{\times}10^{-7}$	$2{\times}10^8$	$1{\times}10^8$	$2{\times}10^4$
1989 — Cray-3 (gallium arsenide)							
$1{\times}10^7$	$1{\times}10^7$	64	$6{\times}10^{-10}$	$6{\times}10^{-10}$	$1{\times}10^{11}$	$6{\times}10^8$	$1{\times}10^4$

A3 | *The Outer Limits of Computation*

Mind without Machine?

In Chapter 4 I suggested that a mind is a pattern that can be impressed on many different kinds of body or storage medium. I went further to say that a mind could be represented by any one of an infinite class of radically different patterns that were equivalent only in a certain abstract mathematical sense. A person's subjective experiences are an abstract property shared by all patterns in this class, so the person would feel the same regardless of which pattern she was instantiated in. This leads to the question that if a mind is ultimately a mathematical abstraction, why does it require a physical form at all? Don't mathematical properties exist even if they don't happen to be written down anywhere? Doesn't the billionth digit of pi exist even if we haven't yet managed to compute it? In the same sense, don't the abstract mathematical relationships that are the feelings of a person exist even in the absence of any particular hardware to compute them? This happens to be an old philosophical conundrum; and though I think it must be true, I do not see how to draw any meaningful conclusions from it, since it seems to imply that everything possible exists. So, instead, let's consider a slightly less sweeping line of thought.

Suppose that a program describing a person is written in a static medium like a book. A superintelligent being who reads and understands the program should be able to reason out the future development of the encoded person in a variety of possible situations. Existence in the thoughts of an intelligent beholder is fundamentally no different than existence in a computer simulation, and we have already suggested that a mind can be satisfactorily encoded in a computer. But an intelligent being can do more with the simulation than simply carry it out rigidly—it need not model accurately every single detail of the beheld and may well choose to skip the boring parts, to jump to conclusions that are obvious to it, to approximate other steps, and to lump together alternatives it does not choose to distinguish. Human authors of fiction do this every day as they create adventures for the characters in their books—our superintelligent being is different only in that its imagination works at such a level of detail that its simulated people are fully real. Like an author of fiction, the being can think in a time-reversed way; it may choose a conclusion and then reason backwards, deciding what must have preceded it. Perhaps the superintelligent being prefers to imagine certain kinds of situations and contrives to maneuver its mental simulations to make them happen. By failing to flesh out unimportant details in the simulation and steering events toward particular conclusions, our intelligent being may create enough peculiarities in the simulated world to attract the notice of the simulated person.

Quantum mechanics, a cornerstone of modern physics, seems to imply that in the real world as we know it, unobserved events happen in all possible ways (another way of saying no decision is made as to which possibility occurs), and the superposition of all these possibilities itself has observable effects, including mysterious coincidences at remote times and places. Is there any connection between these ideas? Once again I have argued myself into a conundrum, though this time one that has some possibility of being answered eventually.

Nondeterministic Thinking

The cellular automata universes featured in Chapter 6 have a clockwork rigidity that makes them easy to think about. But modern physics has revealed much more interesting foundations for our own universe. Just what it all really means is still a matter for fascinating speculation; the only consensus is that the truth is *very* weird. The following is a little self-indulgence—an attempt to harness the strangeness to a practical end. Of course reality will turn out to be far stranger.

Before computers, serious scientific calculations were all difficult, lengthy, and error prone. Practical engineering problems were approached in an ad hoc manner with numerical tables, graphical methods, and clever devices like slide rules and, in this century, mechanical calculators. Automatic computers dramatically changed the situation. Although they could handle astronomically large problems, they required, in advance, an absolutely precise and detailed step-by-step specification of the procedure. Writing such programs, and getting them right, was slow and tedious (though not nearly as slow as doing the calculations themselves by hand). It was a great advantage if the same program could be used for many problems of the same type, and often such generalization made the structure of the calculation

I Think, Therefore I Am
A simulated Decartes correctly deduces his own existence. It makes no difference just who or what is doing the simulation—the simulated world is complete in itself.

clearer. It also spawned the new mathematical field of *computational complexity*, the study of the intrinsic difficulty of computing answers to different kinds of problem.

A general program can handle problems of different sizes, and usually its running time will grow as its problem size is increased. The simplest sorting programs, for instance, take times proportional to the square of the number of items to be sorted. Doubling the number of items quadruples the sorting time. More complicated programs have been discovered whose times grow more slowly, and students of computational complexity were able to prove that the fastest possible sorting programs would require a number of steps proportional to the number of items times its logarithm—sorting one million items should take about about ten thousand times as long as sorting one thousand items. The time grows faster than the number of items but not as fast as its square. The difficulty of other kinds of problem was shown to grow as the cube of the size, or fourth, or some other power. Problems whose solution can be computed in times proportional to (or less than) some fixed power of the problem size are said to be of polynomial time, or P, type. In the computer era they are considered easy.

Another important class of computations turned out to be much harder. One example is the so-called *traveling-salesman* problem, which involves finding the shortest path that passes through each of a given set of cities exactly once. The best exact solutions that anyone has found require that the program check almost every possible route, then pick the best one. The possible routes are enumerated by generating all permutations of the cities. Each permutation itself can be checked in polynomial time, but the number of permutations grows exponentially, multiplying manyfold each time a single city is added. The problem becomes astronomically difficult for quite modest numbers of cities. A hypothetical computer able to explore all possible paths simultaneously (a mere mathematical abstraction known as a *nondeterministic* machine because it does not make up its mind at branchpoints but splits into two machines and goes both ways) could, in principle, solve the problem in polynomial time. For that reason these hard problems are called NP, for nondeterministic polynomial. It turns out that many NP problems can be mathematically converted into one another, and that a fast (P) solution for one would solve all. Alas, no fast solution that works on a real, deterministic computer has been found, and it is not known if such a solution exists. This so called P=NP? question is a central one because some extremely important problems, for instance in the design of optimal hardware and software, and automatic reasoning, are NP. Since exact solutions are much, much too slow, present design systems limp along with approximate methods that do not guarantee the best answer, and sometimes do quite badly.

Let's suppose, as seems likely, that there is no shortcut to NP problems. No matter how fast the conventional machine we use to solve them, small increases in problem size will swamp it. An intelligence designing improvements for itself will encounter many NP issues. The efficiency of its designs, and thus its future, will depend directly on how well it can handle these problems, so heroic methods are warranted.

Replication and Quantum Nondeterminism

A nondeterministic computer, able to spawn indefinitely many versions of itself to examine alternative answers, is a mathematical fabrication. It can be approximated, to a limited extent, by a multiprocessor that contains a fixed number of distinct processors. At each branchpoint in the computation a new processor is invoked. Eventually, however, the last processor will be occupied, and subsequent branches must be evaluated sequentially. The computation can be sped up only by as many times as there are individual processors, making hardly a dent in the astronomical growth rate of NP problems. But what if the number of processors could somehow be increased as the problem size grew?

The ability to reproduce is a sine qua non for the members of any race that aspires to immortality. Our machines do it now, though biological humans are an essential part of the process. As the future portrayed in this book dawns, more and more of the steps will be taken over by machines, until fully automatic manufacture of automatic manufacturing machines is the norm. In the superintelligent future, reproduction of small thinking units will be child's play. It is thus reasonable to imagine small computers that can both compute and make copies of themselves. Reproduction eventually is limited by the finite amount of available material and energy, but the limits could be made astronomically high by breeding very small machines in a very large, energetic nutrient medium (bacteria-sized machines in the oceans of Jupiter, or in pulsar-pumped interstellar dust clouds, let's say!). One such computer would reproduce, to become two, the two would become four, the four eight, the eight sixteen, and so on, in an exponential growth to astronomical numbers matching, up to a point, the exponential growth of an NP problem. If, in one "generation" time step, a machine can do either a certain amount of computation or reproduce itself once, the best strategy would be to reproduce like mad until there are as many machines as there are alternatives to examine, then for each resulting machine to examine one alternative answer. Having computed a possible answer, the machines would then engage in a "mine's bigger than yours" tournament with nearby machines. Two machines would compare each other's answers, and then both would adopt the better of the two results and go off to compete with other machines. The best answer will rapidly spread to all the machines, and any one can then be harvested to learn the final result. In this process, the reproduction phase takes a time linearly proportional to the size of the problem; working out the individual answers is polynomial in the size, and the answer tournament phase, like the reproduction, is linear in the size, allowing moderately sized NP problems to be solved in polynomial time. But very large problems would swamp any amount of available space and would thus still be out of reach. (Earth's biosphere, of course, is performing a computation of just this kind.)

Truly nondeterministic computers are a mathematical fiction. Or are they? Quantum mechanics, a cornerstone of modern physics, has indeterminism at its heart and soul. Outcome probabilities in quantum mechanics are predicted by summing up all the indistinguishable ways an event might happen, then squaring the result. One strange consequence of this is that some otherwise possible outcomes are ruled out by the existence of other possibilities. An excellent example is the *two-slit* experiment. Photons of light radiate from a pinpoint source to a screen broken by two slits, as in the figure on page 184. Those that make it through the slits encounter an array of photon detectors (often a photographic film, but the example is clearer if we use individual sensors that click when struck by light). If the light source is so dim that only one photon is released at a time, the sensors register individually—sometimes this one, sometimes that one. Each photon lands in exactly one place. But, if a count is kept of how many photons have landed on each detector, an unexpected pattern emerges. Some detectors see no photons at all, while ones close to them on either side register many, and a little farther away there is again a dearth. If the number of clicks from each detector is charted, the result is identical to the banded interference pattern that would occur if steady waves had been emitted from the location of the light source and passed through the two slits, to be split there into two waves that interfered with each other at the screen, constructively at some places, destructively at others, as in the figure on page 185.

Two-Slit Experiment
A photon picked up by a detector at screen S might have come through slit A or through slit B—there is no way to distinguish. In quantum mechanics the "amplitudes" for the two cases must be added. At some points on the screen they add constructively, making it likely that a photon will end up there; at nearby points the amplitudes cancel, and no photons are ever found.

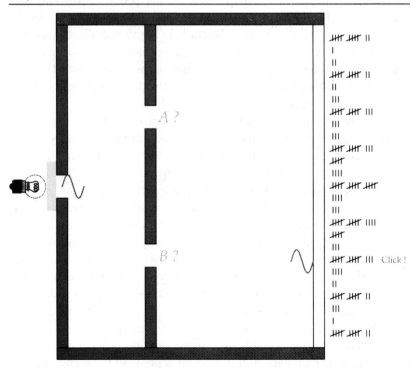

But waves of *what*? Each photon starts from one place and lands in one place; isn't it at just one place on every part of its flight? Doesn't it go through one slit or the other? If so, how does the presence of the other slit *prevent* it from landing at a certain place on the screen? For if one slit is blocked, the total number of photons landing on the screen is halved, but the interference pattern vanishes, and some locations that received no photons with both slits open begin to register hits. Quantum mechanics' answer is that during the flight the position of the photon is unknown and must be modeled by a complex valued wave describing all its possible locations. This ghostly wave passes through *both* slits (though it describes the position of only a single photon) and interferes with itself at the screen, canceling at some points. There the wave makes up its mind, and the photon appears in just one of its possible locations. This wave condition of the photon

Two Slits and Waves
When sound waves are passed through the two slits, an interference pattern results. But no individual clicks are heard. Each wave gently affects all the detectors.

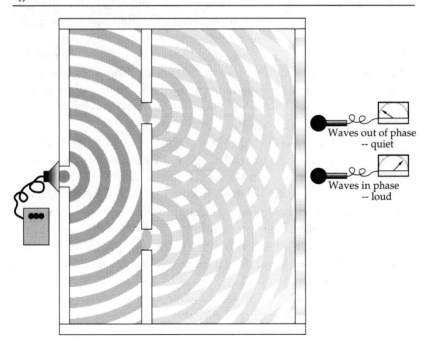

Waves out of phase
-- quiet

Waves in phase
-- loud

before it hits the screen is called a *mixed state* or a *superposition of states.* The sudden appearance of the photon in only one detector is called the *collapse of the wave function.*

This explanation profoundly disturbed some of the same physicists who had helped to formulate the theory, notably Albert Einstein and Erwin Schrödinger. To formalize their intuitive objections, they constructed thought experiments that gave unlikely results according to the theory. In some, a measurement made at one site caused the instant collapse of a wave function at a remote location—an effect faster than light. In another, more frivolous, example, called Schrödinger's Cat, a radioactive decay that may or may not take place in a sealed box causes (or fails to cause) the death of a cat also in the box. Schrödinger considered absurd the theory's description of the unopened box as a mixed state superimposing a live and a dead cat. He suggested that the theory merely expressed ignorance on the part of an observer: in the box the cat's fate was unambiguous. This is called a *hidden-variables* theory; that is, the system has a definite state at all times, but some parts of it are temporarily hidden from some observers.

The joke is on the critics. Many of the most "absurd" thought experimental results have been observed in mind-boggling actuality in clever (and very modern) experiments carried out by Alain Aspect at the University of Paris. These demonstrations rule out the simplest and most natural hidden variables theories, *local* ones, in which, for instance, the hidden information about which slit the photon went through is contained in the photon itself, or ones in which the state of health of Schrödinger's cat is part of the feline.

Nonlocal hidden-variables theories, where the unmeasured information is distributed over an extended space, *are* a possibility. It is easy to construct theories of this kind that give results identical with ordinary quantum mechanics. Most physicists find them uninteresting: why introduce a more complicated explanation with extra variables when the current, simpler equations suffice? Philosophically, also, global hidden-variables theories are only slightly less puzzling than raw quantum mechanics. What does it mean that the "exact position" of a particle is spread out over a large chunk of space? This question was the subject of a lively controversy among the founders of quantum mechanics in the early part of this century. It has recently become of widespread interest again.

Many Worlds and the Doomsday NP Computer

In the two-slit experiment, a photon destined for the screen *might* go through slit A or it might traverse slit B. The interference pattern suggests it somehow manages to do both at the same time. In 1957 Hugh Everett at Princeton published in his PhD thesis what may be the most profligate nonlocal hidden-variables explanation of this puzzle. In Everett's model, the photon does go through both slits, *in different universes.* At each decision point the entire universe, or at least the immediate portion of it, splits into several, like multiple pages from a copying machine. Until a measurement is made, the different versions of the universe lie in close proximity and interfere with each other, causing banded patterns on screens, for instance. A measurement that can distinguish one possibility from another causes the universes to diverge (alternatively, the divergence is the definition of "measurement"). The interference stops, and in each, now separate, universe a different version of the experimenter can contemplate a different unambiguous result.

The word "astronomical" hardly begins to capture the number of distinct universes created every instant under this idea. Such numbers alone may make the thought unappealing, yet we should not be intimidated by mere scale. Shock also greeted the first suggestions of the number of atoms in a speck of matter, or the distance to the nearest stars, or the size and age of the universe. Practical physicists object to the profligacy of universes for a different reason. Once a measurement *is* made, the universes in which the result was other than our own no longer influence us. Postulating their continued existence is an unnecessary complication. Thus stripped, the many-worlds idea is reduced to conventional quantum mechanics. The unnecessary hidden variables (that identify *which* universe you are talking about) are removed.

Can quantum mechanical indeterminism help solve difficult problems? So-called holographic methods have been demonstrated that use coherent laser light to simultaneously search for certain patterns (fingerprints, or enemy radar signals, for instance) in large fields of other patterns. For the right problem, holographic methods are fast and efficient, and the action can be interpreted as the effect of mixed states. It can also be viewed in a more classical way as light-wave interference. Waves are interesting because their spread is a kind of reproduction, and different parts of a waveform can be used to perform different parts of a computation. But as yet only linear speedups have been proposed or achieved. NP problems remain difficult.

But the many-worlds idea has other consequences. John Gribbin is a writer and physicist who has expanded on its more bizarre possibilities in several stories, articles, and books. One of his stories has the following plot.

The Doomsday Device

Two builders of a future super (immensely expensive) particle accelerator have a problem. The machine has been completed for months, but so far has failed on each attempt to use it. The problem is not in the design but seemingly just in the designer's bad luck. Lightning caused a power outage just at turn on, or a fuse blew, or a janitor tripped over a cable, or a little earthquake triggered an emergency cutoff; each incident was different, and apparently unrelated to the others.

But perhaps the failures are an enormous stroke of luck. New calculations suggest that the machine is powerful enough to trigger a collapse of the vacuum to a lower energy state. A cosmic explosion might radiate out at the speed of light from the accelerator's collision point, eventually destroying the entire universe. What a close call!

Or was it? If the universe had been destroyed, there would be no one left to lament the fact. What if the many-worlds idea were correct? In some universes the machine would have worked. For all practical purposes those worlds would have ceased to exist. Only in the remainder would a pair of puzzled physicists be scratching their heads, wondering what had gone wrong this time. Given so many nearly identical universes, the destruction of a few seems of small consequence. An idea strikes them. Why not reinforce the weak points in the machine so that a random failure within it is extremely unlikely, then wire it to a detector of a nuclear attack, like the doomsday machine in Stanley Kubrick's film Dr. Strangelove? *An attack would be met by the destruction of the offending universe. Only those universes in which the attack had not happened, for some reason (the commanding general had a heart attack, the missile launch system failed, the premier had a fit of compassion...), would live to wonder about yet another close call. The machine in* Strangelove *was ineffective as a deterrent unless the other side was aware of it. Not so the many-worlds version. No attack (that anyone will notice) can occur so long as it operates, no matter how secret its existence.*

Preventing nuclear war is a laudable objective perhaps worth the destruction of untold numbers of possible universes. But can we use the same approach to solve day-to-day problems? Getting back to the NP problem, could parallel universes somehow be used to search alternative possible solutions simultaneously, allowing one conventional computer to act like a nondeterministic one? Let's begin by wiring a universe-destroying device to a computer, so that it can be activated by a certain computer instruction. While we are at it, let's also connect a true random number generator, based on a hiss from a hot resistor or clicks from a geiger counter. Potential solutions to NP problems are characterized by a number, for instance the total path length in the traveling-salesman problem, that we seek to minimize. Guess a probable length, and write a program to choose a path at random (using the generator you added). Have the program trigger the doomsday device if the randomly generated path is longer than your guess. Run the program. Most likely you will find that the random path your program has generated is shorter than your guess, because universes in which a longer path came up are wiped out. But suppose there *is no path* that short. Well, then your computer must have broken somehow and failed to destroy the universe.

It is a nuisance to have your computer break in an uncontrolled way, so why not provide an easily fixed weak link? Install another thermally or radioactively controlled device that, with low probability, can interrupt your computer's calculation. If you run your program now, it will either produce an answer shorter than your guess, or else be interrupted before it finishes. In the former case, choose a shorter path length; in the latter, choose a longer one. Then run the program again. Keep doing this until you find a path length L such that the machine finds an answer with a guess of L, but interrupts itself when the guess is L-1. The L-length solution is an optimal answer to your problem. The search for a proper L can, of course, be incorporated into the program itself to make the process fully automatic. An optimal strategy for doing this, called a *binary search*, can finish the search in a number of steps proportional to the logarithm of L, a mere 20 steps when L is a million. Each step, which involves generating each random answer, takes polynomial time. So the answer is yes, a many-worlds version of quantum mechanics can be used to solve arbitrarily hard NP problems in modest times.

The cost in destroyed universes is staggering, however. In the nuclear-war-preventing use of the doomsday device, if the annual chance of an attack were 50%, about half of the possible universes would be destroyed each year. But in a traveling-salesman problem with 100 cities only a few solutions out of 100^{100} (a 1 followed by 200 zeros) will be optimum, so a doomsday computer solving the problem would destroy about 100^{100} worlds for each one it let survive! The growth rate of new universes is much larger than that, so perhaps it does not matter.

Only one component of this solution to the NP problem cannot be immediately constructed today, and that is the doomsday trigger itself. Gribbin's design depends on highly speculative physics that has lost favor as of this writing. But is it really necessary to destroy the whole universe if the answer comes out wrong? Of course not. Subjectively, it is just as good to merely destroy yourself! Now *that* outcome can be achieved today. Wire your computer's doomsday connection to a cranial explosive charge, for instance. If you run the traveling-salesman program you will, with overwhelming odds, blow your brains out. But it's quick and should be painless. And, in one world out of 100^{100}, you will, by a tremendous stroke of luck, survive *and* have the right answer. Your cranial explosive will be intact, ready to solve the next problem.

The above idea would work with other problems of life. If an outcome you desire, however unlikely, fails to materialize, destroy yourself. In some universe, you will survive, having won your bet. If the many-worlds interpretation of quantum mechanics is correct, as it might be, why isn't suicide a common solution to everyday problems? The demographics of multiple universes may have some bearing on the answer. If you are careless about losing your life, there will be fewer copies of you among the universes. A universe picked at random will contain mostly individuals who successfully struggle to avoid death whenever possible. If the universes are truly infinite in number, this has little meaning—one trillionth of infinity is still the same infinity. The consequences are serious, however, if the number is merely huge. And if not all possible outcomes are pursued, then destroying yourself with high probability could, in fact, truly end your existence.

One World, Not Many?

It may be that the indeterminate nature of quantum mechanics is simply a kind of illusion and that there is only one world. Here is an outline of a model where the uncertainties at any location, or the hidden variables, are simply "noise" from the rest of the universe.

Imagine, somewhere, there is a spherical volume uniformly filled with a gas made up of a huge but finite number of particles in motion. Pressure waves pass through the gas, propagating at its speed of sound, s, and suppose no faster signal can be sent. The sphere has resonances that correspond to wave trains passing through its entire volume at different angles and frequencies. Each combination of a particular direction and frequency is called a *wave mode*. There is a mathematical transformation called the (spatial) *Fourier transform* that arranges these wave modes very neatly and powerfully. The Fourier transform combines the pattern of pressures found over the original volume of the sphere (V) in various ways to produce a new spherical set of values (F). At the center of F is a number representing the average density of V. Immediately surrounding it are (complex) numbers giving the intensity of waves, in various directions, whose wavelength just spans the diameter of V. Twice as far from the center of F are found the intensities of wave modes with two cycles across V; these are surrounded by another shell containing modes whose wavelength is one third the diameter of V, and so on. Each point in F describes a wave filling V with a direction and a number of cycles given by the point's orientation and distance from the center of F. Another way of saying this is: direction in F corresponds to direction in V; radius in F is proportional to frequency in V. Since each wave is made of periodic clusterings of gas particles, the interparticle spacing sets a lower bound on the wavelength, thus an upper bound on frequency, and a limit on the radius of the F sphere. The closer the particles, the larger F must be.

A theorem about Fourier transforms states that if sufficiently high frequencies are included, then F contains about as many points as V has particles, and all the information required to reconstruct V is found in F. In fact, F and V are simply alternative descriptions of the same thing, with the interesting property that every particle in V

contributes to the value of each point in F, and every point in F is reflected as a component of motion of every particle of V.

If the particles in V bump into one another, or interact in some other nonlinear way, then energy can be transferred from one wave mode to another—that is, one point in F can become stronger at the expense of another. There will be a certain amount of random transference among all wave modes. Besides this, there will be a more systematic interaction between "nearby" wave modes—those very similar in frequency and orientation, thus near each other in the F space. In V, such waves will be in step for large fractions of their length. Because the gas is nonlinear, the periodic bunching of gas particles caused by one mode will influence the bunching ability of a neighboring mode with a similar period.

So nearby points in F interact systematically, distant points do not. The interaction can be considered a physics of the F world. If the physics is rich enough, it may be able to support the basis of complex structures, life and intelligence, just as does ours. Imagine a physicist made of F stuff, for whom points in F are simply locations, not complicated functions of another space. We can deduce some of the "laws of physics" this inhabitant of F will find by reasoning about effects in V, and translating back to F. In the following list, such reasoning is in italics:

- **Dimensionality:** F is three dimensional. *If V is three dimensional, each wave train will be described by its orientation, given by two angles, say azimuth and elevation, and by its frequency. Frequency in V becomes radius in F, while the two angles in V remain angles in F. If V were an n dimensional sphere, F would also have n dimensions.*

- **Locality:** Points near to each other in F can exchange energy in consistent, predictable ways while distant points cannot. *Two wave trains in V that are very similar in direction and frequency are in step for a long portion of their length, and the nonlinear bunching effects will be roughly the same cycle after cycle along the length. Wave modes distant from one another, on the other hand, whose crests and troughs are not correlated, will lose here, and gain there, and in general appear like mere random buffetings to each other.*

- **Interaction Speed:** There is a characteristic speed at each point in F. Points far away from the center of F interact more quickly than those closer in. *An interaction is the nonrandom transfer of energy from one wave mode to another. The smallest repeated unit in a wavetrain is a cycle. An effect which happens in a similar way at each cycle can have a consistent effect on a whole wave train. Effects in V propagate at the speed of sound, so a whole cycle can be affected in the time it takes sound to traverse it (this is the time period of the wave). The outermost parts of F correspond to wave modes with the highest frequencies, and thus the fastest interaction rates.*

- **Uncertainty Principle:** The energy of a point in F cannot be determined precisely in a short time. The best accuracy possible improves linearly with duration of the measurement. *The energy at a point in F is the total energy of a particular wavetrain that spans the entire volume V. As no signal in V can travel faster than the speed of sound, discovering the total energy in a wavetrain would involve waiting for signals to arrive from all over V, a time much longer than the basic interaction time. In a short time, the summation is necessarily over a proportionately small volume. Since the observer in F is itself distributed over V, exactly which smaller volume is not defined—and thus the measurement is uncertain. As the time and the summation volume increase all the possible sums converge to the average, and the uncertainty decreases.*

- **Superposition of States:** Most interactions in F will appear to be the sum of many possible ways the interaction might have happened. *When two nearby wavetrains interact, they do so initially on a cycle-by-cycle basis, since information from distant parts of the wavetrain arrives only at the speed of sound. Each cycle contains a little energy from the wavetrain in question and a lot of energy from many other waves of different frequency and orientation passing through the same volume. This "background noise" will be different from one cycle to the next along the wavetrain, so the interaction at each cycle will be slightly different. When all is said and done—that is, if the information from the entire wavetrain is collected—the total interaction can be interpreted as the sum of the cycle-by-cycle interactions. Sometimes energy will be transferred one way by one cycle and the opposite way by a distant one, so the alternatives can cancel as well enhance one another.*

These and other properties of the F world contain some of the strangest features of quantum mechanics but are the consequence only of an unusual way of looking at a prosaic situation. There are a few differences. The superposition of states is statistical, rather than a perfect sum over all possibilities, as in traditional quantum mechanics. This makes only a very subtle difference if V is very large but might result in a very tiny amount of "noise" in measurements that could help distinguish the F mechanism from other explanations of quantum mechanics.

The model as presented does not exhibit the effects of special relativity in any obvious way, and this is a serious defect, if we hope to wrestle it into a description of our world. There is something wrong in the way it treats time. It does have one property that mimics the temporal effects of a general relativistic gravitational field. Time near the center of F runs more slowly than at the extremes, since the interactions are based on lower frequency waves. At the very center, time is stopped. The central point of F never changes from its "average density of the whole sphere" value, and so is effectively frozen in time. In general relativity the regions around a gravitating body have a similar property: time flows slower as one gets closer. Near very dense masses (that is, black holes), time stops altogether at a certain distance.

A few of modern physics' more exotic theories have a possible explanation in this model. Although energy mainly flows between wave modes very similar in frequency and direction (such as points adjacent in F), nonlinearities in the V medium should permit some energy to flow systematically between harmonically related wave modes, for instance between one mode and another on the same direction, but twice as high in frequency. Such modes of energy flow in F provide "degrees of freedom" in addition to the three provided by nearby points. They can be interpreted, when viewed on the small scale, as extra dimensions (energy can move this way, that way, that way and also *that* way, and *that* way...). Since a circumnavigation from harmonic to harmonic will cover the available space in fewer steps than a move along adjacent wave modes, these extra dimensions will appear to have a much smaller extent than the basic three. The greater the energy involved, the more harmonics may be activated and the higher the dimensionality. Most physical theories these days have tightly looped extra dimensions to provide a geometric explanation for the basic forces. Ten and eleven dimensions are popular, and new forces suggested by some theories may introduce more. If something like the F explanation of apparent higher dimensionality is correct, there is a bonus. Viewed on a large scale, the harmonic "dimensions" are actual links between distant regions of space and, properly exploited, could allow instantaneous communication and travel over enormous distances.

Big Waves

Now, forget the possible implications of the F idea as a mechanism for quantum mechanics and consider our universe on the grand scale. It is permeated by a background of microwave radiation with a wavelength centered around 1 millimeter, a length slowly increasing as the universe expands and cools. It affects, and is affected by, clouds of matter in interstellar space and thus interacts with itself nonlinearly. If we do a universe-wide spatial Fourier transform of this radiation (that is, treat *our* world as V), we end up with an F space with properties much like those of the previous section. The expansion of the universe adds a new twist. As our universe gradually expands, the wavelengths of the background radiation increase. As the wavelengths get longer and longer, the relative rate of time flow in the F world slows down. Any inhabitants of F would be ideally situated to practice the "live forever by going slower and slower as it gets colder and colder" strategy proposed by Freeman Dyson. By now they would be moving quite slowly—their fastest particle interactions would take several trillionths of a second. In the past, shortly after the big bang, when the universe was dense and hot, the F world would have been a lively place, running millions or billions of times faster. In the earliest moments of the universe, the speed would have been astronomically faster.

The first microsecond of the big bang could represent eons of subjective time in F. Perhaps enough time for intelligence to evolve, realize its situation, and seed smaller but eventually faster life in the V space. Though on the large scale F and V are the same thing, manipulation of one from the other, or even communication between the two, would be extraordinarily difficult. Any local event in either space would be diffused to nondetectability in the other. Only massive, universe-spanning projects with long-range order would work, and these would take huge amounts of time because of the speed limits in either universe. Real-time interaction between V and

F is ruled out. Such projects, however, could affect many locations in the other space as easily (in many cases more easily) as one, and these could appear as entropy-violating "miracles" there. If I lived in F and wanted to visit V, I would engineer such a miracle that would condense a robot surrogate of myself in V, then later another miracle that would read out the robot's memories back into an F-accessible form.

The Fourier transform that converts V into F is identical except for a minus sign to the inverse transform that converts the other way. Given just the two descriptions, it would not be clear which was the "original" world. In fact, the Fourier transform is but one of an infinite class of "orthogonal transforms" that have the same basic properties. Each of these is capable of taking a description of a volume and operating over it to produce a different description with the same information, but with each original point spread to every location in the result. This leads to the possibility of an infinity of universes, each a different combination of the same underlying stuff, each exhibiting quantum mechanical behavior but otherwise having its own unique physics, each oblivious of the others sharing its space. I don't know where to take that idea.

Bibliography

Prologue

Cairns-Smith, A. G. 1982. *Genetic Takeover and the Mineral Origins of Life.* Cambridge: Cambridge University Press.

—— 1985. *Seven Clues to the Origin of Life: A Scientific Detective Story.* Cambridge: Cambridge University Press.

—— 1985 (June). "The First Organisms." *Scientific American* 253:90-98.

Calder, Nigel. 1983. *Timescale: An Atlas of the Fourth Dimension.* New York: Viking Press.

Desmond, Kevin. 1986. *A Timetable of Inventions and Discoveries.* New York: M. Evans.

Sagan, Carl. 1977. *The Dragons of Eden: Speculations on the Evolution of Human Intelligence.* New York: Random House.

Chapter 1: Mind in Motion

Ashby, W. Ross. 1963. *An Introduction to Cybernetics.* New York: John Wiley & Sons.

Asimov, Isaac. 1950. *I, Robot.* New York: Doubleday.

Bakker, Robert T. 1986. *The Dinosaur Heresies.* New York: William Morrow.

Braitenberg, Valentino. 1984. *Vehicles: Experiments in Synthetic Psychology.* Cambridge: MIT Press.

Brooks, J., and G. Shaw. 1973. *Origin and Development of Living Systems.* New York: Academic Press.

Buchsbaum, Ralph. 1948. *Animals without Backbones.* 2nd ed. Chicago: University of Chicago Press.

Feigenbaum, Edward A., and Julian Feldman, eds. 1963. *Computers and Thought.* New York: McGraw-Hill.

Fichtelius, Karl-Erik, and Sverre Sjölander. 1972. *Smarter than Man? Intelligence in Whales, Dolphins and Humans.* New York: Ballantine Books.

Griffin, Donald R. 1984. *Animal Thinking.* Cambridge: Harvard University Press.

Lane, Frank W. 1962. *Kingdom of the Octopus: The Life History of the Cephalopoda.* New York: Pyramid Publications.

Marsh, Peter. 1985. *Robots.* New York: Crescent Books.

Mayr, Ernst. 1982. *The Growth of Biological Thought.* Cambridge: Harvard University Press.

McCorduck, Pamela. 1979. *Machines Who Think.* San Francisco: W. H. Freeman.

Minsky, Marvin, ed. 1985. *Robotics.* New York: Doubleday.

——— 1986. *The Society of Mind.* New York: Simon & Schuster.

Moynihan, Martin. 1985. *Communication and Noncommunication by Cephalopods.* Bloomington: Indiana University Press.

Nilsson, Nils, ed. 1984. *Shakey the Robot: Artificial Intelligence Center Technical Note 323.* Menlo Park: SRI International.

Pawson, Richard. 1985. *The Robot Book.* London: W H Smith & Son.

Pratt, Vernon. 1987. *Thinking Machines: The Evolution of Artificial Intelligence.* Oxford: Basil Blackwell.

Raphael, Bertram. 1976. *The Thinking Computer: Mind inside Matter.* San Francisco: W. H. Freeman.

Reichardt, Jasia. 1978. *Robots: Fact, Fiction and Prediction.* Middlesex: Penguin Books.

Thorson, Gunnar. 1971. *Life in the Sea.* New York: McGraw-Hill.

Walter, W. Grey. 1961. *The Living Brain.* Middlesex: Penguin Books.

Wiener, Norbert. 1965. *Cybernetics, or Control and Communication in the Animal and the Machine.* Cambridge: MIT Press.

Chapter 2: Powering Up

Babbage, Charles. 1961. *Charles Babbage and His Calculating Engines.* Edited by Philip Morrison and Emily Morrison. New York: Dover Publications.

Berkeley, Edmund C. 1949. *Giant Brains or Machines That Think.* New York: John Wiley & Sons.

Booth, Andrew, and Kathleen Booth. 1956. *Automatic Digital Calculators.* London: Butterworths.

Cale, E. G., L. L. Gremillion, and J. L. McKenney. 1979. "Price/performance patterns of U.S. computer systems." *Communications of the ACM* 22(4): 225-230.

Dertouzos, Michael L., and Joel Moses, eds. 1979. *The Computer Age: A Twenty-Year Review.* Cambridge: MIT Press.

Dowling, John E. 1987. *The Retina: An Approachable Part of the Brain.* Cambridge: Harvard University Press.

Drexler, K. Eric. 1986. *Engines of Creation*. New York: Doubleday.

Eames, Charles, and Ray Eames. 1973. *A Computer Perspective*. Cambridge: Harvard University Press.

Hubel, David, ed. 1979 (September). *The Brain—Scientific American*, special issue, 241(3).

Kandel, Eric R. 1976. *Cellular Basis of Behavior: An Introduction to Behavioral Neurobiology*. New York: W. H. Freeman.

Kandel, Eric R., and J. H. Schwartz. 1982. "Molecular basis of memory." *Science* 218:433-436.

Kuffler, Stephen W., and John G. Nichols. 1976. *From Neuron to Brain: A Cellular Approach to the Function of the Nervous System*. Sunderland, Massachusetts: Sinauer Associates.

Lazou, Christopher. 1986. *Supercomputers and Their Use*. Oxford: Clarendon Press.

Minsky, Marvin, and Seymour Papert. 1969. *Perceptrons: An Introduction to Computational Geometry*. Cambridge: MIT Press.

Moreau, Rene. 1984. *The Computer Comes of Age*. Cambridge: MIT Press.

von Neumann, John. 1958. *The Computer and the Brain*. New Haven: Yale University Press.

Queisser, Hans. 1988. *The Conquest of the Microchip*. Cambridge: Harvard University Press.

Randell, Brian, ed. 1973. *The Origins of Digital Computers: Selected Papers*. Berlin: Springer-Verlag.

Rosen, Saul. 1971. *ACM 71: A Quarter-Century View*. New York: Association for Computing Machinery.

Squire, Larry R. 1986. "Mechanisms of memory." *Science* 232:1612-1619.

Turn, Rein. 1974. *Computers in the 1980s*. New York: Columbia University Press.

Weik, Martin H. 1957. *A Second Survey of Domestic Electronic Digital Computing Systems*. Aberdeen, Maryland: Army Ballistic Research Laboratories, report no. 1010.

Wolfe, Jeremy M. 1986. *The Mind's Eye: Readings from Scientific American*. New York: W. H. Freeman.

Wolken, Jerome J. 1975. *Photoprocesses, Photoreceptors, and Evolution*. New York: Academic Press.

Chapter 3: Symbiosis

Boorstin, Daniel J. 1983. *The Discoverers*. New York: Random House.

Time-Life Books, eds. 1985. *Computer Images: Understanding Computers Series*. Arlington: Time-Life Books.

Chapter 4: Grandfather Clause

Forward, Robert L. 1980. *Dragon's Egg.* New York: Ballantine Books.

———— 1984. *The Flight of the Dragonfly.* New York: Simon & Schuster.

Hofstadter, Douglas R., and Daniel C. Dennett, eds. 1981. *The Mind's I.* New York: Basic Books.

Hoyle, Fred, and John Elliot. 1962. *A for Andromeda.* New York: Harper.

Rucker, Rudy. 1983. *Software.* Middlesex: Penguin Books.

Sperry, Roger. 1982. "Some effects of disconnecting the cerebral hemispheres." *Science* 217:1223-1226.

Vinge, Vernor. 1984. *True Names.* New York: Bluejay Books.

Chapter 5: Wildlife

Axelrod, Robert. 1984. *The Evolution of Cooperation.* New York: Basic Books.

Cohen, Fred. 1984. *Computer Viruses.* Los Angeles: University of Southern California.

Dewdney, A. K. 1984 (May), 1985 (March), and 1987 (January). "Core wars." *Scientific American* 250:14-17, 252:14-23, 256:14-18.

Dawkins, Richard. 1976. *The Selfish Gene.* Oxford: Oxford University Press.

———— 1982. *The Extended Phenotype.* Oxford: Oxford University Press.

———— 1986. *The Blind Watchmaker.* New York: W. W. Norton.

Oliver, B. M., and J. Billingham. 1971. *Project Cyclops: A Design Study of a System for Detecting Extraterrestial Intelligent Life.* Moffett Field, California: NASA/Ames Research Center.

Rosen, Eric C. 1981. *Vulnerabilities of Network Control Protocols: An Example.* Cambridge: Bolt Beranek and Newman, Inc.

Sagan, Carl. 1985. *Contact.* New York: Simon & Schuster.

Shoch, John F., and Jon A. Hupp. 1982. "The 'worm' programs: early experience with a distributed computation." *Communications of the ACM* 25(3):172-180.

Thompson, Ken. 1984. "Reflections on trusting trust." *Communications of the ACM* 27(8):761-763.

Chapter 6: Breakout

Barrow, John D., and Frank J. Tipler. 1986. *The Anthropic Cosmological Principle.* Oxford: Oxford University Press.

Dyson, Freeman. 1979. *Disturbing the Universe.* New York: Harper & Row.

———— 1988. *Infinite in all Directions.* New York: Harper & Row.

Gardner, Martin. 1983. *Wheels, Life and Other Mathematical Amusements.* New York: W. H. Freeman.

Hawking, Stephen W. 1988. *A Brief History of Time: From the Big Bang to Black Holes*. New York: Bantam Books.

Weinberg, Steven. 1977. *The First Three Minutes*. New York: Basic Books.

Appendix 3: The Outer Limits of Computation

DeWitt, Bryce S., and Neill Graham, eds. 1973. *The Many-Worlds Interpretation of Quantum Mechanics*. Princeton: Princeton University Press.

Feynman, Richard P., Robert B. Leighton, and Matthew Sands. 1965. *The Feynman Lectures on Physics*. Vol. 3. New York: Addison-Wesley.

Gribbin, John. 1984. *In Search of Schrödinger's Cat: Quantum Physics and Reality*. New York: Bantam Books.

—— 1985. "Doomsday device." *Analog Science Fiction–Science Fact* 105:22-127.

Robot Pals
Some of the author's mechanical collaborators, with the years of the association.

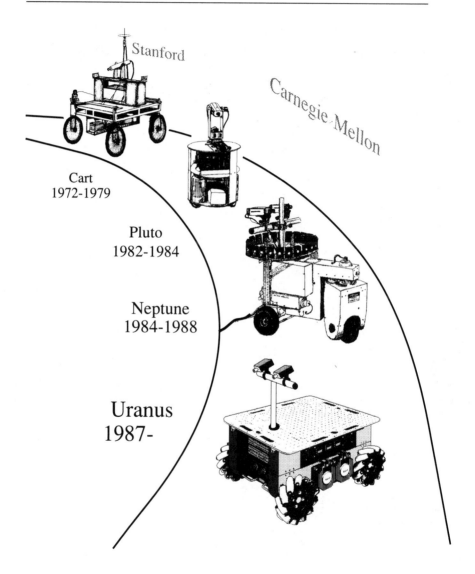

Stanford

Carnegie Mellon

Cart
1972-1979

Pluto
1982-1984

Neptune
1984-1988

Uranus
1987-

Acknowledgments & Illustration Credits

THIS book has roots deep in my childhood. My memory isn't up to the task of acknowledging them individually, but my thanks go to the authors of science and science fiction, teachers, librarians, science fair organizers and friends who helped shape my mental world through four decades. My memory *is* good enough to recall that my younger sisters Elizabeth and Alice were the long-suffering sounding boards for many years of my long-winded speculations.

In late 1971, when I arrived as a graduate student at the Stanford Artificial Intelligence Laboratory, a lively debate was just winding down that had been sparked by a proposal from Dick Fredericksen in his self-published newsletter, *A Word in Edgewise*. Over several articles he had developed the concept of achieving immortality, and much else, by replacing a human nervous system, bit by bit, with a more durable artificial equivalent. The exchange interested me greatly because the idea had occurred to me years before, in high school, but had suffered from lack of a receptive audience. At SAIL, Fredericksen's proposal had polarized those who took it seriously. Bruce Baumgart was its chief proponent, while Larry Tessler found it dehumanizing. My own thoughts about the future of intelligent machinery crystallized in discussions with Rod Brooks, Bruce Bullock, Mike Farmwald, Bob Forward, Don Gennery, Erik Gilbert, Bill Gosper, David Grossman, Brian Harvey, Marc Le Brun, Robert Maas, John McCarthy, Ed Mcguire, Dave Poole, Jeff Rubin, Clem Smith, Russ Taylor, Lowell Wood, and quite a few others. In 1975 I wrote an essay on the subject that evolved over the years into several articles and, eventually, into this book.

The discussions continued when I came to Carnegie Mellon University in 1980. Here it is my pleasure to acknowledge mind-stretching exchanges with Mike Blackwell, Kevin Dowling, Alberto Elfes, Larry Matthies, Pat Muir, Gregg Podnar, Olin Shivers, and Richard Wallace. I would also like to thank the administration of the Robotics Institute, especially Raj Reddy and Takeo Kanade, for maintaining an environment that allows me to pursue long-range goals. I am equally grateful to the Office of Naval Research, and my program director, Alan Meyrowitz, for providing the steady funding that has supported my basic research since 1981.

Though I vaguely intended to develop my ideas to book length in 1975, it was only in 1985 that I seriously began to work on a manuscript. By an amazing coincidence, within two weeks of undertaking the project I received a letter from Howard Boyer, the newly arrived Editor for Science and Medicine at Harvard University Press, inviting me to write just such a book. In the three years since then, Howard has whipped the book through a grueling series of writing and publishing hurdles. I am deeply grateful for his interest and insight.

The first hurdle—to produce a detailed outline that would pass editorial muster at Harvard Press—was surmounted with the aid of extensive reviews written by Vernor Vinge. The many drafts of the manuscript which followed were greatly improved by the Press's referees, whose comments were variously encouraging, informative, and stern. I thank Rod Brooks, Richard Dawkins, Kee Dewdney, Bruce Donald, John Dowling, Bob Forward, John McCarthy, Pamela McCorduck, and others whose identity I have not learned.

The most dramatic improvements in the book occurred when it was put in the hands of my manuscript editor, Susan Wallace. Susan reorganized the text from a ragged collection of ideas into a cohesive whole, setting the stage for— and coaching to completion—a rewrite that made a night-and-day difference in the book's quality.

With exceptions noted below, the line art in the book was drawn by me on a Macintosh II from Apple Computer. The programs used were *Cricket Draw* from Cricket Software and *SuperPaint* from Silicon Beach Software, with occasional dips into digitized clip-art collections—the *McPic!* packages from Magnum Software and *ClickArt* packages from T/Maker Graphics. Some art was scanned from hand drawings and photographs with the *Thunderscan* program and hardware from Thunderware. Earlier versions of many of the drawings had been produced on smaller Macintoshes with *MacPaint*, *MacDraw*, *FullPaint*, and *MacDraft*.

Mike Blackwell redrew "Intelligence on Earth" (page 18) in *Cricket Draw* from my *MacDraw* original. "The Retina" (page 54) was drawn on a Macintosh SE by Mary Jo Dowling, using *Adobe Illustrator* from Adobe Systems. She worked from an illustration which appeared in *The Retina* by John Dowling (no relation). The gear and integrated circuit icons in "A Century of Computing" (page 64) were drawn by Gregg Podnar using *MacPaint*. "A Robot Bush" (page 103) was produced by a program which I wrote with help from Mike Blackwell, in Apple Computer's *MPW C* language, running on a Macintosh II. It contains one quarter of a million line segments and took ten hours to compute. The "Selfish Martians" cartoon (page 142) was drawn by Kimberlee Faught with *Cricket Draw*. The pictures of the Cart, Pluto, and Neptune in

"Robot Pals" (page 202) are digitized and touched-up renderings of pencil drawings by Bill Nee. The picture of Uranus is a digitized photograph touched up by me and Gregg Podnar.

The following organizations provided photographs and granted permission to reproduce them: "Walking Machines" (page 27), courtesy of Odetics, Inc.; "Three Fingers" (page 30), courtesy of David Lampe, MIT; "Autonomous Navigation" (page 33), courtesy of Denning Mobile Robotics, Inc.; "Object Finding" (page 35), courtesy of SRI International; "The Retina" (page 54), courtesy of John Dowling; "ENIAC" (page 76), © Smithsonian Institution; "Magic Glasses" (page 87), courtesy of United Technologies/Hamilton Standard; "Robot Proxy" (page 88), courtesy of Naval Ocean Systems Center; "Unreal Estate—The Road to Point Reyes" (page 92), © 1986 Pixar.

The book was designed by Joyce C. Weston, Marianne Perlak, and Mike Blackwell. It was typeset by Mike Blackwell in the TEX document preparation system created at Stanford University by Don Knuth, as instantiated in the *Textures* program by Addison-Wesley, running on a Macintosh II. The typeface is Palatino, obtained from Adobe Systems. Camera-ready copy was generated on a Linotronic 300 digital typesetter owned by the Robotics Institute of Carnegie Mellon University, driven by a Macintosh SE. Early drafts of the book were printed on Apple laserwriters.

Index